Advances in Intelligent Systems and Computing

Volume 333

Series editor

Janusz Kacprzyk, Polish Academy of Sciences, Warsaw, Poland
e-mail: kacprzyk@ibspan.waw.pl

About this Series

The series "Advances in Intelligent Systems and Computing" contains publications on theory, applications, and design methods of Intelligent Systems and Intelligent Computing. Virtually all disciplines such as engineering, natural sciences, computer and information science, ICT, economics, business, e-commerce, environment, healthcare, life science are covered. The list of topics spans all the areas of modern intelligent systems and computing.

The publications within "Advances in Intelligent Systems and Computing" are primarily textbooks and proceedings of important conferences, symposia and congresses. They cover significant recent developments in the field, both of a foundational and applicable character. An important characteristic feature of the series is the short publication time and world-wide distribution. This permits a rapid and broad dissemination of research results.

Advisory Board

More information about this series at http://www.springer.com/series/11156

Katharina Kinder-Kurlanda
Céline Ehrwein Nihan
Editors

Ubiquitous Computing in the Workplace

What Ethical Issues? An Interdisciplinary
Perspective

 Springer

Editors
Katharina Kinder-Kurlanda
GESIS—Leibniz Institute for the Social
  Sciences
Cologne
Germany

Céline Ehrwein Nihan
University of Applied Sciences in Business
  and Engineering Vaud (HEIG-VD)
Yverdon-les-Bains
Switzerland

ISSN 2194-5357   ISSN 2194-5365 (electronic)
Advances in Intelligent Systems and Computing
ISBN 978-3-319-13451-2   ISBN 978-3-319-13452-9 (eBook)
DOI 10.1007/978-3-319-13452-9

Library of Congress Control Number: 2014956486

Springer Cham Heidelberg New York Dordrecht London

Printed on acid-free paper

Springer International Publishing AG Switzerland is part of Springer Science+Business Media
(www.springer.com)

Preface

Information and Communication Technology (ICT) has undergone a great evolution in the last decades. The miniaturization of computers and the development of ever smaller and more powerful sensors have permitted the emergence of more embedded and smarter computing, which has started to profoundly change our habits and daily life [1]. The ICT research efforts presented in the first part of this volume (Voirin and De Paz et al.) give a first indication of the potential of such ubiquitous computing (ubicomp) to realize the vision of an 'internet of things' connecting physical and digital artifacts. Some applications are already on the market, others are about to appear, and others have not yet been imagined.

In this volume, we put our attention on the developments that are taking place in the workplace. It is obvious that ubicomp has also begun to enter our working environments and to transform our conceptions of work, our relation to it, as well as our professional skills and well-being. And these evolutions are going to continue.

Yet managers and HR professionals seem to be mostly unaware of these developments. They are relatively poorly prepared for the organizational and managerial transformations that come with ubicomp in workplaces. While some authors have started to question the implications of ubicomp for Information Technology Management in companies [2], made studies on the acceptance of Intelligent Working Environments (IWE) [3], or tried to sketch scenarios of the future workplace [4], almost none have proposed to examine the ethical and managerial issues of these developments.

This observation has led us to create a research group in order to address this specific question. Over the last two years there have been three workshops on the topic of Human Resource Management Challenges of Intelligent Working Environments (IWE&HRM), organized by an international program committee and taking place at the HEIG-VD in Yverdon-les-Bains, Switzerland. These workshops brought together experts from various disciplines who have been working on the socio-ethical impact of technological developments in ubicomp, pervasive computing, and ambient intelligence in various research and development projects. The aim was to gather the distributed expertise and experiences of these experts within different application settings in order to formulate recommendations for (human resource) managers who are increasingly faced with new developments in workplaces becoming "intelligent" that they may not fully comprehend, yet need to be prepared for.

The workshops were elaborately designed and prepared (establishment of a program, clear-cut objectives, and instructions). Participants were requested to submit work in advance as a basis for the discussion. One or two participants would also be invited to present a paper related to their individual research in order to gain momentum for the discussion. Each workshop resulted in a report. During the workshops particular attention was paid to the interdisciplinarity of the discussion (e.g., speaking slots), the achievements of the predefined objectives, and the integration of the points of agreement/disagreement in the workshop reports.

In large part, the papers presented in this volume directly result from these meetings, in particular form the third one which more specifically focused on socio-ethical concerns and was part of a research conference on "Ubiquitous Computing in the Workplace: What Ethical Issues?" in Yverdon-les-Bains in 2013.[1] At the end of the last workshop, it appeared that the papers that had been presented (De Paz et al., Wiegerling, Kinder-Kurlanda et al.) as well as the ensuing discussion (Ehrwein Nihan) deserved to be published—all the more so in view of the scarcity of publications related to the topic. We decided to expand the volume with three additional invited papers (Voirin, Hilty, Poltier). All authors were selected on the basis of their expertise as well as on their previous peer-reviewed work in the field. Contributions were reviewed by two members of the program committee and proposed changes were implemented as requested.

The book has been conceived and structured in two parts as follows: A first part is dedicated to the current developments of ubicomp systems designed for the workplace. Guy Voirin, from the Swiss Center for Electronics and Micro-technology (CSEM), and Francisco De Paz, Sara Rodríguez, Carolina Zato, and Juan M. Corchado from the Department of Computer Science and Automation in Salamanca (Spain) offer us an introductory insight into potential fixed as well as wearable IWEs.

The second part of the volume focuses on the ethical issues raised by the current developments of ubicomp in the workplace. These are considered from multi-level and cross-disciplinary perspectives. In a first step, the philosopher Klaus Wiegerling from the Kaiserslautern Technical University in Germany proposes a meta-ethical investigation of the challenges raised by the current developments in ubiquitous computing. He particularly addresses questions of the personal identity of the subject who has to act within an ambient intelligent environment, of his/her perception of the life-word ("Lebenswelt"), and of the possibility of his/her choice.

We then turn toward more specific socio-ethical perspectives: first, the ICT scientist Lorenz Hilty proposes to look at the evolution of the discourses between the first ethical reflections on ICT in the 1970s and on the progressive appearance of

---

[1] http://www.heig-vd.ch/campus/evenements/research-conference-ubiquitous-computing-in-the-workplace-what-ethical-issues.

ubicomp over the last two decades. He highlights the persistence of certain issues while bringing out the emergence of new aspects. The anthropologist Katharina Kinder-Kurlanda and the management, technology, and economics researcher Daniel Boos also explore this socio-ethical level, but in a less socio-historical and more socio-anthropological way. Relating experiences from two ubicomp projects, they show how these specific examples are connected to wider societal trends and ethical issues of informational ubiquity such as the requirement for more transparency and information control.

The next paper, dedicated to the results of the IWE&HRM research project (see above), takes an even closer look at the applied ethical issues raised by the development of ubicomp in the workplace. After general conceptual and socio-ethical considerations, the ethicist Céline Ehrwein Nihan examines some of the concrete impacts that these environments might have for both employers and employees. On this basis she then makes some suggestions regarding the rules that should be respected in order to favor an adequate implementation of ubicomp in the workplace.

The volume ends with a paper by the philosopher Hughes Poltier, who again widens the scope of analysis with an ethico-political perspective on the discussion. Highlighting the power at stake in every technical device, the author focuses on the risk of ubicomp development increasing the power imbalance within organizations and more generally in society. By doing so, he reminds us that we are all responsible to engage in discussion and decision making regarding the design and implementation of future ubicomp technology.

Acknowledgments

We would like to express our sincere thanks to Tiziana Boni, who helped us to improve the English of some of the papers as well as all to those who contributed to this project.

References

1. Dourish, P., Bell, G.: Divining a Digital Future: Mess and Mythology in Ubiquitous Computing. MIT Press, Cambridge Ma/London (2011)
2. Patten, K., Passerini, K.: From Personal Area Network to Ubiquitous Computing: Preparing from a Paradigm Shift in the Workplace. In: Wireless Telecommunications Symposium, pp. 225–233. Institute of Electrical and Electronics Engineers (2005)
3. Röcker, C.: Acceptance of Future Workplace Systems: How the Social Situation Influence the Usage Intention of Ambient Intelligence Technologies in Work Environments. In: Proceeding of the 9th International Conference on Work with Computer System (WWCS 09), CD-ROM (2009)
4. Bühler, C.: Ambient Intelligence in Working Environments. In: Stephanidis, C. (ed.) Universal Access in HCI, Part II, pp. 143–149. Springer, Berlin/Heidelberg (2009)

Logo HES-SO

Hes·so

Haute Ecole Spécialisée
de Suisse occidentale

Fachhochschule Westschweiz

University of Applied Sciences and Arts
Western Switzerland

Logo HEIG-VD

**HAUTE ÉCOLE
D'INGÉNIERIE ET DE GESTION
DU CANTON DE VAUD**

www.heig-vd.ch

Organizing Committee

Bernard Baertschi	University of Geneva
Tiziana Boni	University of Applied Sciences in Business and Engineering Vaud
Daniel Boos	Independent Researcher
Silna Borter	University of Applied Sciences in Business and Engineering Vaud
Zarina Charlesworth	University of Applied Sciences in Business Neuchâtel
Florian Dufour	University of Applied Sciences in Business and Engineering Vaud
Céline Ehrwein Nihan	University of Applied Sciences in Business and Engineering Vaud
Laurence Firoben	University of Applied Sciences in Business and Engineering Vaud
Esther Ivanyi	University of Applied Sciences in Business and Engineering Vaud
Katharina Kinder-Kurlanda	GESIS, Leibniz Institute for the Social Sciences
Mario Konishi	University of Applied Sciences in Business and Engineering Vaud

Contents

Part I
Present Ubiquitous Computing
Developments in the Workplace

RETRACTED CHAPTER: An Integrated System for Helping Disabled and Dependent People: AGALZ, AZTECA, and MOVI-MAS Projects

Juan F. De Paz, Sara Rodríguez, Carolina Zato and Juan M. Corchado

Abstract This article presents three successful case studies oriented to disabled and dependent people. These three case studies are the results of the following corresponding projects: Autonomous aGent for monitoring ALZheimer's patients (AGALZ), which facilitates the monitoring and tracking of patients with Alzheimer's; AZTECA, which is formed by a set of tools that facilitate the work of disabled people in their work environment; and MOVI-MAS, which simulates a 3D work environment enabling the detection of dangerous situations. These tools were developed using an agent platform called PANGEA, which is a platform to develop open multi-agent systems, specifically those including organizational aspects such as virtual agent organizations.

Keywords Disabled people · Dependent people · Organization of agents · Ambient intelligence · Work environment

1 Introduction

Due to the advance of technologies and communications, intelligent systems have become an integral part of many people's lives. Available products and services have become more varied and capable, and users expect to be able to personalize a product or service to meet their individual needs, no longer accepting a "one size

The original version of this chapter was retracted: The retraction note to this chapter is available at http://doi.org/10.1007/978-3-319-13452-9_8

J.F. De Paz (✉) · S. Rodríguez · C. Zato · J.M. Corchado
Departamento Informática Y Automática, Universidad de Salamanca, Salamanca, Spain
e-mail: fcofds@usal.es

S. Rodríguez
e-mail: srg@usal.es

C. Zato
e-mail: carol_zato@usal.es

J.M. Corchado
e-mail: corchado@usal.es

© Springer International Publishing Switzerland 2015, corrected publication 2024
K. Kinder-Kurlanda and C. Ehrwein Nihan (eds.), *Ubiquitous Computing in the Workplace*, Advances in Intelligent Systems and Computing 333,
DOI 10.1007/978-3-319-13452-9_1

fits all" solution. Personalization can range from simple cosmetic factors, such as custom ring tones, to the complex tailoring of the presentation of a shopping Web site, to a user's personal interests and their previous purchasing behavior [1–3]. These innovative techniques are expected to expand to a wide range of fields. One of the segments of the population expected to benefit from the advance of personalized systems, which will contribute to improve their quality of life [4], is people with disabilities [5, 6]. It is important to study the different procedures that facilitate the adaption of these systems to the disability of each user, allowing them to experience improvement in their quality of life and in their work production.

There are currently a number of barriers that make it difficult for people with disabilities to be incorporated into the workforce and, consequently, for businesses to include them among their personnel. The greatest challenges for incorporating these individuals into the workforce are personal autonomy (mobility), information processing (language, knowledge of numbers, learning tasks, spatial orientation), attitude toward work (responsibility, attention, rhythm, organization, work relationships, security, interest…), emotional control, interpersonal relationships, and self-determination. It becomes necessary, therefore, to provide new tools that can eliminate these barriers and facilitate the integration of this group of individuals into the workforce. The solutions that can make it possible to reach these goals should consider the type and degree of disability, since the objectives for the integration of these individuals are conditioned by the special needs of each type of disability.

In the near future, public and private companies will be provided with intelligent systems specifically designed to facilitate interaction with human users. These intelligent systems will be able to personalize the services offered to the users according to their specific profile. It is necessary to improve the services provided, as well as the way they are offered [7]. Technologies such as multi-agent systems (MAS) [8] and Ambient Intelligence (AmI) [9, 10], which are based on mobile devices, have been recently explored as systems of interaction with dependent people [11]. These systems can provide support in the daily lives of dependent people [12].

The main purpose of AmI systems when they first appeared was to improve human–environment interaction by means of computerized pervasive environments. The research community rapidly became aware of the potential of such systems to provide individual users with specialized solutions, particularly elderly people and people with disabilities, who can undoubtedly benefit from many innovative services related to health care, location, work environments, security, etc.

AmI-based systems aim to improve the quality of life, offering more efficient and user-friendly services and communication tools to interact with other people, systems, and environments. It seems reasonable to assume that elderly and dependent people are the segments of the population that would benefit the most from the development of these systems [5].

At present, there is an increasing need to supply constant care and support to sectors of the society such as the elderly and dependent disabled people [13]. Consequently, the drive to find more effective ways to provide such care has become a major challenge for the scientific community.

In Europe, the Eurostat Agency has established that 78 % of the severely disabled aged people 16–64 do not work and only 16 % of those who face work restrictions are provided with some assistance at work [14]. In Spain, the INE Institute obtained the following results: Of the 1,171,900 people with certified disabilities in 2010, aged from 16 to 64 years, only 423,700 had an active job [15]. The Office for Disability Issues in the UK found that the 46.3 % of disabled people are employed compared to 76.2 % of non-disabled people [16].

The creation of secure, unobtrusive, and adaptable environments for monitoring and optimizing health care will become vital. Some authors [13] consider that tomorrow's healthcare institutions will be equipped with intelligent systems capable of interacting with humans.

This article presents two projects specially oriented toward disabled people in the workplace and the PANGEA multi-agent platform, which enables the integration of these projects.

The remainder of the paper is structured as follows: The next section introduces the chosen approach related to multi-agent systems. Sections 3–5 present the corresponding projects AZTECA, Autonomous aGent for monitoring ALZheimer's patients (AGALZ), and MOVI-MAS. Finally, in the last section, some conclusions are drawn.

2 The Multi-agent Architecture Approach

The BISITE research group of the University of Salamanca has carried out several projects and research related to healthcare environments and workplace integration [17–19]. MAS and other architectures based on intelligent devices have recently been explored as supervision systems for medical care for dependent people [20–23]. These intelligent systems aim to support dependent persons in all aspects of daily life, predicting potentially dangerous situations and delivering physical and cognitive support.

At present, there is a clear tendency within the new generation of software applications to favor autonomy, robustness, flexibility, and adaptability of the developed systems. The multi-agent approach [24] has become increasingly relevant for developing distributed and dynamic intelligent environments and therefore fulfils the requirements and goals of AmI and Ambient Assisted Living. In this sense, it is important to integrate intelligent and dynamic mechanisms to learn from past experiences and therefore provide users with better tools for supplying health care.

Three of the most difficult aspects when dealing with AmI systems are as follows: (i) heterogeneity of hardware and, therefore, of the agents responsible for system control (ii) scalability, since new devices or services should be able to be easily added as the system grows, and (iii) high dynamicity, since the agents that control and process the information enter and exit the system continually.

These features and the high interaction between agents have led to the evolution from multi-agent systems to virtual organizations of agents (VOs). VOs [25–27] are a means of understanding system models from a sociological perspective. A VO is an open system formed by the grouping and collaboration of heterogeneous entities and includes a separation between form and function that requires defining how a behavior will take place. Furthermore, VO encompasses the following concepts:

- Roles: These enable the representation of agent functionalities and a separation from the rest of its cognitive abilities.
- Norms: These are vital when managing information and interaction between agents. They enable defining and imitating agent behaviors with high precision.
- Organizational topology: This allows the development of software models that are independent while still able to work together. According to the requirements of each system, agents move in and out of the various VOs that compose the system.

The main problem of implementing a VO is the lack of platforms that support such systems. The main function of an agent platform is to provide a runtime environment for the agents. Given the nature of the presented projects and case studies, in our examination of the literature, no platform was found on which the three systems can be developed; therefore, a generic platform called PANGEA was created in order to allow the easy development and subsequent integration of the three systems.

2.1 PANGEA Platform

The main novelty of this platform lies in its design, which is oriented toward providing the services required to create agents that offer services that can be integrated to achieve a global system. The detailed operation of the platform can be found at [28, 29].

One of the greatest advantages of this platform is the communication platform which, using the IRC standard, offers a robust and widely tested communication system that can handle a large number of connections and ensure scalability and reliability. Another reason that justifies the scalability of the platform is the way it models the services, that is, as services inside the agents or as SOA architecture compliant services using Web services. The platform offers an IDE, which facilitates the implementation process. It automatically offers the skeleton of an agent, and the communication between agents can be implemented with few lines of code. Finally, the platform admits mobile agents and agents in any programming language; it is not necessary to learn a new language in order to use it.

In general, this is an innovative platform. No other platform containing these characteristics, such as the ability to integrate different services with high heterogeneity or different functional natures, is known to exist. The main platform agents are shown in Fig. 1.

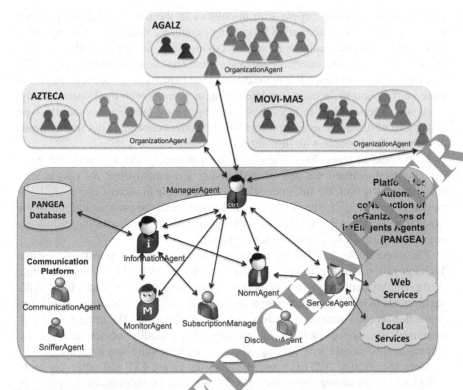

Fig. 1 Overview of the systems

As discussed below, this platform has enabled the development of an integral multi-agent architecture creating a novel system to provide services aimed at integrating people with visual, hearing, and motor skill impairments.

3 The AZTECA Project

3.1 Motivation

An important trend in contemporary societies is the rapid development of ICT, which has influenced the lives of people in every way; another is the sensitivity that exists in governments, companies, and associations toward enabling people with disabilities or at risk of exclusion to lead an independent life, which includes a very relevant ability to acquire gainful employment. While the effective integration of people with disabilities in the work place is a huge challenge for societies, it also presents an opportunity to make use of ICT.

Currently, there are numerous barriers that keep people with disabilities from joining the labor market and companies from incorporating them into the workplace.

Challenges of including people with disabilities in the labor market are personal autonomy (mobility), information processing (language, numeracy, learning tasks, or spatial orientation), attitude to work (responsibility, attention, rhythm, organization, labor relations, safety, or interest), emotional control, interpersonal relationships, and other factors. It is therefore essential to provide new tools to overcome these barriers and to facilitate the integration of persons with disabilities into the workforce. The solutions to meet these challenges must take into account the type and degree of disability, as labor integration objectives depend on the special needs of each group of people with disabilities.

In the field of technology, there have been several recent advances that have significantly facilitated the task of daily living and employment for people with disabilities. However, the full integration of these people in society, and in the labor market in particular, continues to be a challenge. One of the main limitations of the existing proposals is that they focus on very specific problems and, in general, are oriented toward a single type of disability.

The scope of the AZTECA project (Intelligent Environments with Accessible Technology for Work) is to research and develop new technologies that contribute to the employment of groups of people with visual, hearing, and motor skills impairments in office work environments through a service architecture oriented to interaction, communication, mobility, and self-employment. The project has revolutionized the interoperability, accessibility, and usability of services for the employment of people with hearing, visual, and motor skills impairments.

More specifically, AZTECA specifically aims to use CIT and other key technologies to create an intelligent work environment that provides integral support for the needs of persons with disabilities for their integration into the workforce. AZTECA promotes the entrance of people with disabilities into the labor market and their development of an independent life. In addition, the proposed set of tools can be used by businesses to facilitate integration into the workforce, helping companies to comply with the Spanish Organic Law 13/1982, of April 7, Social Integration of Disabled (LISMI), which establishes the obligation of public and private companies employing more than 50 workers to hire no fewer than 2 % of their workforce as persons with disabilities.

For these reasons, we propose the development of an innovative architecture that integrates both location and identification services, as well as communication and training tools as interactive services that provide adaptive and personalized access.

1.2 System Overview

AZTECA is composed of a series of tools that comprise a global system. These tools include registration services, rapid writing services, adaptation to alternative peripheral devices, adaptive interfaces, virtual interpretation of sign language, active learning services with TV, and location services. Three of these tools are presented below.

3.2.1 Proximity Activation System

The entry port to the system includes a proximity activation system that detects the user and configures the work environment. The proximity activation system is based on the detection of presence using the ZigBee technology [30, 31]. Every computer in the room must have a ZigBee router assigned, and the system has to know the exact position of the user at every moment. Furthermore, all users have to carry a ZigBee tag, which is responsible for identifying each individual. Once the ZigBee tag carried by the person has been detected and identified, its location is delimited within the proximity of the sensor that identified it (Fig. 2).

This system is able to personalize the workspace to improve the adaptation to the company workflow. Whatever disability the person may have, this type of individual adaptation allows the workplace to be adapted automatically, facilitating work productivity and removing the existing barriers, such as an inability to turn on the computer with the proximity detection system.

The system is designed as a VO where several agents are involved. Every disabled user in the proposed system carries a ZigBee tag, which is detected by a ZigBee Reader Agent located in each system terminal and is in continuous communication with the Client Computer Agent. Thus, when a user tag is sufficiently close to a specific terminal (within a range defined according to the strength of the signal), the ZigBee Reader Agent can detect the user tag and immediately send a message to the Client Computer Agent, which is coordinated by the ZigBee Coordinator Agent. The system uses a local area network (LAN) infrastructure with a wake-on-LAN protocol for the remote switching on and off of the equipment.

This tool works together with the customization tool, also displayed as a sub-organization within the PANGEA platform. The detection and identification of a user make it possible to detect any special needs and for the computer to be automatically adapted for individual use. This allows the system to define and

Fig. 2 User carrying a tag and tag located on the table to detect the RSSI signal

manage the different profiles of people with disabilities, facilitating their job assimilation by automatically switching on or off the computer upon detecting the user's presence or initiating a procedure that automatically adapts the computer to the personal needs of the user.

Because the system uses a LAN infrastructure, the wake-on-LAN protocol is used for switching on the computers. Wake-on-LAN/WAN protocol is a technology that allows a computer to be turned on remotely by a software call. It can be implemented in both LAN and wide area networks (WANs) [32]. It has many uses, including turning on a Web/FTP server, remotely accessing files stored on a machine, telecommuting, and in this case, turning on a computer even when the user's computer is turned off [33].

3.2.2 Translation Tool

The translation tool emerged as a result of the difficulty encountered by employers in communicating with hearing-impaired employees. Given the ineffectiveness of avatar translators, the solution chosen was to study the most important communication needs and to provide some recorded videos with commands and explanations specifically related to the performance of a particular job. These videos are pre-recorded by a sign language interpreter and stored on a Web server where they can be accessed anytime through the request of an issuing agent. The issuing agents, deployed on both smartphones (android o iPhone) and computers, will be responsible for playing the video required at that moment. Receptor agents, also available for smartphones and computers, will be responsible for capturing, either by text or by voice, the information or instruction that the employer wishes to transmit to the disabled employee.

The translator agent, which is called Video Translator Agent, is deployed in the platform. It is responsible for receiving the instruction and mapping the specific video used by the emitter agent who is requesting the transfer.

The translator agent is deployed within the sub-organization Translator Organization. Within this organization, a translation tool designed for people with visual disabilities is also displayed. The tool consists of a vibrating bracelet that receives impulses to transmit messages in Morse code. As with the previous system, the employer sends a text through a mobile agent or an agent deployed on his/her computer. The text is received by a second translator agent, Morse Translator Agent, which interprets the message by translating it into Morse code and then uses Bluetooth to send the message content to the bracelet. The bracelet transmits vibrations to the disabled worker, who is the receiver of the message. These interactions are shown in Fig. 3.

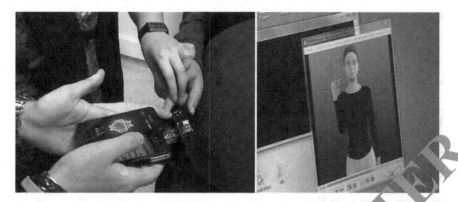

Fig. 3 Mobile phone, chip, and vibrator used for the translator tool and avatar shown on the computer

3.2.3 Learning and Monitoring TV

This tool was also developed with the intention of facilitating the learning skills of disabled people and to develop their work. Learning is done through the use of the TV at home. The TV and remote control were chosen as communication tools primarily due to the great familiarity that users have with these devices and their management. TV is any easy device to use, even for those who do not hold any particular technical knowledge.

The current version is based on a Web platform, designed for use in low-resource hardware support devices (such as set-top box or Raspberry Pi). Moreover, it seeks to create an ecosystem that allows knowing the state of the home and its inhabitants. In order to accomplish this, we will use the information provided by various sensors located in the home.

The display of the TV is achieved by embedding a viewer plugin in the browser, which provides access to a Freeview tuner, either integrated or external (usually connected via USB). Thus, no bandwidth Internet connection is necessary, nor is an expensive Internet connection required to use the system.

In order to transmit the information from the disabled user to the control center, the system may use the user's Internet connection; however, these connections are usually expensive and slow in rural environments. In order to reduce cost and to allow transmitting the information, the system may be integrated with a WIFI network with a low cost.

Through this system, users can check the tasks to be performed, notices to be sent to work and even receive notifications or reminders and schedules. This system is interconnected to the translator tool so the user can receive notifications in different ways (Fig. 4).

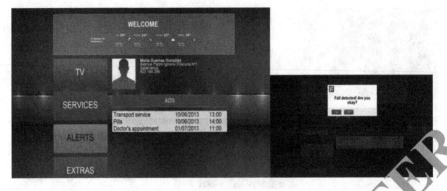

Fig. 4 Screenshots of the TV

4 The AGALZ Project

4.1 Motivation

The number of elderly people in the European Union is increasing. It is estimated that in 15 years, this group will comprise one-third of the population of the EU and creating systems that facilitate monitoring and tracking of dependent persons seems beneficial [34]. This situation is not unique to the EU as it can also be seen in other industrialized countries. For example, it is estimated that by 2020, one-sixth of the US population will be 65 years or older [35]. Dependent persons often need care and tracking which in the early stages of dependency can be carried out with correct patient monitoring. The situation tends to worsen for 20 % of seniors 85 years or older who require constant monitoring, care and, in some cases, specialized medical attention or hospitalization.

This situation has led us to propose the creation of a monitoring and tracking system which is both unobtrusive and able to adapt to changes in the environment. Additionally, tracking patients must allow attending personnel to care for patients' medical needs. In other words, medical personnel, who are monitoring patients, must know the state of the patients' treatment and care. Sensors and readers can be used to track patients and determine their location, as in the case of those with specific disabilities such as Alzheimer's. These patients are easily disoriented, which makes it necessary to locate them and to activate an alert in the case of an emergency to avoid situations such as the patient exiting the building.

The idea to create the AGALZ project in combination with ALZheimer's Multi-Agent System (ALZMAS) evolved from a desire to solve this type of problem and to facilitate the process of monitoring and tracking patients in care homes for the elderly.

4.2 System Overview

The ALZMAS project integrates a set of agents, such as the AGALZ agent, that manage the daily tasks of personnel working in a senior care facility. Figure 5 shows the system agents and how they communicate with each other.

The system is composed of the following agents:

- Patient: This agent is in charge of managing the information for the users in the facility. Each patient is associated with an agent responsible for monitoring the user information and for ensuring that everything is in order. The monitored information includes the patient's location, his/her treatment, and daily tasks that should be carried out. The agent is also responsible for knowing the location of all patients and sending an alarm upon detecting that a patient is located in an area where they should not be. The agent stores the patients' movements throughout the day for future analysis.
- Manager: This agent is in charge of managing the alarms that are activated by the system and managing the tasks that must be carried out by medical personnel in the hospital. It manages the treatments that must be administered by the doctors and the list of tasks required to attend to the patients. Tasks are assigned to patients through the AGALZ agent; this is done dynamically and in execution time, taking into account the profile, abilities, previous time required to complete tasks, and the total and temporary workload for each patient. The system takes into account the time frame and time constraints of each task in order to correctly assign tasks and to avoid delay. This system is also capable of rescheduling tasks according to the state of completion fulfilled by personnel. If the system determines that a nurse is unable to complete a task for a patient, the task is reassigned to another care provider. The system also allows the care providers to alert the system of a possible delay, thus allowing the reassignment of tasks to others. This agent works with the manager agent to monitor tasks.
- Doctor: There is a doctor agent for each doctor. This agent is executed on the doctor's mobile device and is in charge of managing the tasks performed by the

Fig. 5 Agents involved

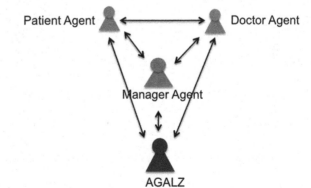

Patient Agent

Doctor Agent

Manager Agent

AGALZ

doctor. The agent communicates with the manager agent to update the task list and with the AGALZ agent to analyze patient progress.

- AGALZ Schedule: This agent is executed on the nurse's mobile device. The agent recovers the task list for a specific nurse and is responsible for ensuring monitoring the state of the tasks. If the system determines that it is impossible to complete treatment, it alerts the manager agent to reassign the task. There are various reasons for being unable to carry out a task, from an emergency situation to equipment failure. The agent ensures that the nurse's tasks are rescheduled according to the indicated time frame and constraints. In order to reassign tasks, the agent uses the profile information of each nurse, previous time required complete tasks, and the necessary resources (Fig. 6).

The system includes some hardware components that monitor the state of the patients. The facility has installed sensors to determine the location of the patients [36]. A passive RFID [37, 38] at 125 KHZ is used to facilitate this task. When a patient is detected outside of a secure area, the agent manager is alerted so that it may in turn alert one of the nurses currently online. The main hardware components are as follows:

- Bracelet: The bracelets contain the tag which identifies the patients and the doctors. The tag worn by the users is shown in Fig. 7.
- Reader: The readers are located in the doors of the facility and in other common walkways. The readers have a range of 2 m. They are connected to computers,

Fig. 6 Screen of the application

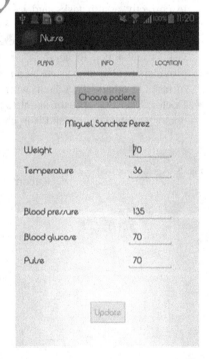

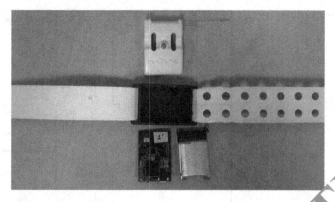

Fig. 7 RFID tag in *bracelet*

Fig. 8 RFID reader

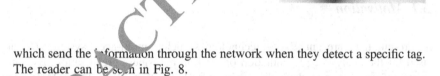

which send the information through the network when they detect a specific tag. The reader can be seen in Fig. 8.

A schematic diagram of the system operation is shown in Fig. 9. Each of the nurses and doctors has a mobile device which receives the information of the tasks they can perform. The devices are connected to the central computer, which is where the manager agent and a patient agent for each patient are executed. The manager agent is in charge of managing the system information. The RFID reader is connected to the central computer and sends patient location information to the computer where the patient agent is executed. This agent is then responsible for informing the manager agent if needed.

Fig. 9 Overview of the system

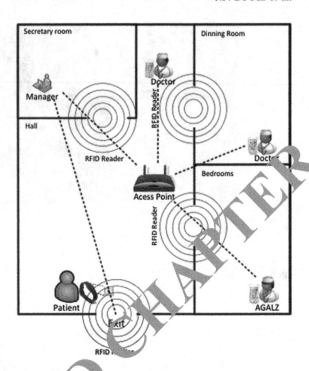

5 The MOVI-MAS Project

5.1 Motivation

At present, there are a number of technologies, such as multi-agent systems, that are used for the analysis and simulation of work in a business organization. By gathering information and through continuous observation, these technologies can identify the requirements of employment, the procedures of the company, the services available in the workplace, and their physical configuration.

The contribution of agent-based computing to the field of computer simulation mediated by agent-based simulation (ABS) provides benefits such as methods for the evaluation and visualization of multi-agent systems or training future users of the system [39]. Many new technical systems are distributed systems that involve complex interactions between humans and machines, which notably reduces their usability. The properties of ABS make it especially suitable for simulating this kind of system. The idea is to model the behavior of human users in terms of software agents. However, it is necessary to define new middleware solutions that allow the connection of ABS simulation and visualization software.

The goal of the MOVI-MAS project has been to design and develop a tool that allows to give a three-dimensional view of all the information gathered from a workplace, in order to analyze its efficiency, make decisions on improving processes, and display the simulated environment of the organization.

The developed system simulates the required information and indicates how it will produce the work development process. It is also possible to analyze and predict the behavior and evolution of the work environment for people with disabilities who need an environment suited to their abilities.

The basis of this project is the integration of 3D simulation techniques and intelligent agents. The combination of both techniques allows the simulation and visualization of activities in a working environment (Fig. 10).

5.2 System Overview

In order to test the system, a Multi-agent System (MOVIMAS) is developed to simulate an office environment and study the problems of accessibility experienced by people with disabilities in performing different jobs. The MAS is designed as a VO [40] modeled after reality. All workers, jobs, and interaction elements, such as architectural barriers, are modeled as agents; these are then grouped into departments according to their availability and their occupation (Human Resources Department, Quality Department, Production Department with the Costumer Service, and Mail sub-departments).

Fig. 10 MAS and 3D simulation techniques integration

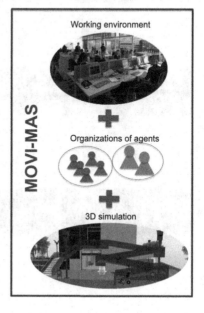

As can be seen in the following diagram, the application from the user's point of view is composed of

- User. As in most applications, the user is one of the most important elements. The user is the main actor of the application and will be in continuous interaction with the 3D visualization tool through its interface.
- 3D visualization tool. It is responsible for communication with the agent platform providing the necessary data.
- Agent platform. It communicates with the 3D visualization tool to send required data (Fig. 11).

The agent platform and the visualization tool are connected by a middleware called Middleware Infrastructure to Simulate Intelligent Agents (MISIAs, [41], created specifically for this purpose. MISIA connects with the visualization tool developed with the Unity 3D engine [40] that simulates office activities in 3D. It also includes other features such as the ability to create and delete agents or to configure different architectural barriers from the 3D simulation application.

Figure 12 illustrates an example of using the system and shows the interaction of agents in REPAST and 3D application. MISIA allows the simulation, visualization, and analysis of agent behavior. MISIA makes use of technologies for the development of well-known and widely used MAS and combines them so that it is

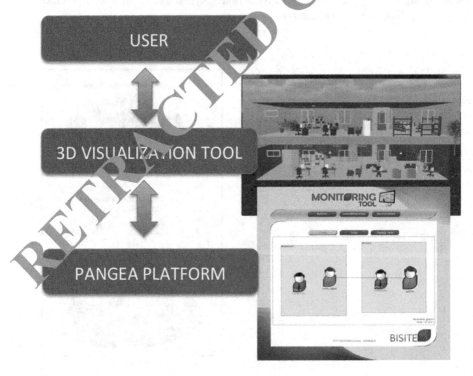

Fig. 11 Elements of the MOVIMAS system

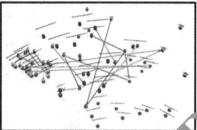

Fig. 12 MOVI-MAS editor and the agents involved on the *right*

possible to use their capabilities to build highly complex and dynamic systems. Moreover, MISIA presents a reformulation of the FIPA [42] protocol and holds several advantages such as independence between the model and the visualization components, improvement of the visualization components, which makes it possible to use the concept of "time" essential for the simulation and the analysis of the behavior of agents, and improvements to user capabilities, including the addition of several tools such as message visualization, 2D (and 3D agents), behavioral analysis, and statistics [25].

The main purpose of MOVIMAS is to search for the optimal working conditions of the employees in the office, thus allowing greater efficiency. For this purpose, several simulations of the tasks that workers have to perform and a 3D simulation application will represent and determine the degree of success of employees in their work. There may frequently occur unusual cases in the simulations, such as a person who needs a wheelchair and cannot access the top floor of the office because the elevator is broken and no ramps are enabled, or a worker who takes a long time to perform certain tasks because the floor contains a step and accessing the destination may require a longer detour. For added versatility in the simulations, and because the application was not dependent on the plan and the disposition of the office, the system uses algorithm A* [43] to search for the shortest path. Thus, represented agents are able to find the optimal path for the tasks that have to perform.

The case study was modeled as a MAS, making it possible to study, at a low-level, all the interactions that agents have with their environment and then to analyze and visualize the results in Repast in order to predict results after many simulations. Thus, given an initial configuration for the VO agents, it is possible to predict what the optimal disposition for the work environment is. The three-dimensional simulation of the office environment here is a great incentive to make the visualization more versatile and accurate and to provide a much more interactive interface for users of the simulation application.

Figure 13 represents the structure of the VO. DA, DC, and DT agents are the heads of department and sub-departments (Customer Service, Quality Control, and

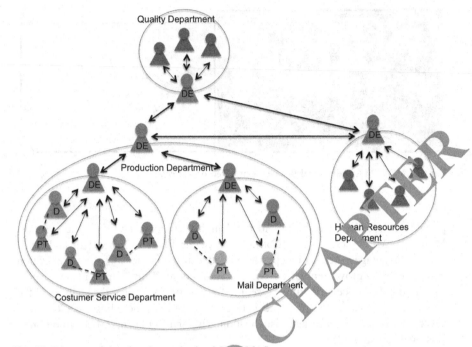

Fig. 13 Diagram of the virtual organization MO_ MAS

Production, respectively). The Mail Department belongs directly to the Production Department. Communication is carried out directly between the agents of the Human Resource and Quality Department and the workers, and this is done directly between them without going through the department heads. The purpose of this diagram is to represent the system like a VO with a federation topology [44], where agents relinquish some of their autonomy and only communicate with their superior, which is precisely what occurs in reality.

One of the most important features of social simulations is that they make it possible to easily observe emergent behavior. Realistic simulations with a significant level of detail, although complex, are best suited to represent processes that study or want to obtain an explanation of the processes or predict outcomes (Fig. 4). MOVIMAS encourages the use of complex simulations for study and enables the analysis, simulation, and visualization of both system interactions and the results obtained in a multi-agent behavior. Furthermore, the interactions between agents are well defined with the use of FIPA protocols and supported VO.

Fig. 14 3D visualization example

6 Conclusions

This collection of systems is specifically oriented toward facilitating the integration of people with disabilities into the workplace to assist them in their daily lives and to help dependent people to improve their quality of life as well. These systems were conceived after extensive research based on the needs of such users. The systems were developed as independent projects with some different funding; however, because we believe that they are closely related, we chose to build an integral system and ultimately developed the new PANGEA platform. Thanks to the PANGEA platform, the system can be easily designed and deployed since the platform itself provides agents and tools for the control and management of any kinds of open MAS or VO. Moreover, the platform makes it possible to deploy different agents, even those included in mobile devices, and to communicate with the agents embedded in the different sensors.

Acknowledgments The AZTECA project has been supported by the Spanish CDTI Proyecto de Cooperación Interempresas IDI-20110343, IDI-20110344, IDI-20110345 Project supported by FEDER funds. The AGALZ project work has been partially supported by the Spanish Ministry of Science project "THOMAS" (TIN2006-14630-C03-03) and the IMSERSO 137/07. The MOVI-MAS system has been developed under the Spanish Ministry of Science project "OVAMAH" (TIN2009-13839-C03) and was also supported by Cátedras Indra-Fundación Adecco.

References

1. Pluke, M., Petersen, F., Brown, W.: Personalization and user profile management for public internet access points (PIAPs). CiteSeerX Scientific Literature Digital Library. Online Resource (2009). doi:10.1.1.117.8111
2. Corchado, J.M., Bajo, J., De Paz, J.F., Rodríguez, S.: An execution time neural-CBR guidance assistant. Neurocomputing **72**(13–15), 2743–2753 (2009)

3. Bajo, J., Corchado, J.M., Rodríguez, S.: Intelligent guidance and suggestions using case-based planning. In: Weber, R.O., Richter, M.M. (eds.) Case-Based Reasoning Research and Development, vol. 4626, pp. 389–403 (2007)

4. Corchado, J.M., Bajo, J., Abraham. A.: GERAmI: Improving the Delivery of Healthcare. IEEE Intel. Syst. **23**(2), 19–25 (2008)

5. Carretero, N., Bermejo, A.B.: Inteligencia Ambiental. Centro de Difusión de Tecnologías, Universidad Politécnica de Madrid, España, CEDITEC (2005)

6. Fraile, J.A., Bajo, J., Corchado, J.M., Abraham, A.: Applying wearable solutions in dependent environments. IEEE Trans. Inf. Technol. Biomed. **14**(6), 1459–1467 (2010)

7. Macarro, A., Bajo, J., Jiménez, A., De la Prieta, F., Corchado, J.M.: Learning system to facilitate integration through lightweight devices. In: Proceedings of FUSION, Chicago, USA (2011)

8. Bajo, J., De Paz, J.F., Rodríguez, S., González, A.: Multi-agent System to Monitor Oceanic Environments. J. Integr. Comput. Aided Eng. **17**(2), 131–144 (2010)

9. Tapia, D.I., Fraile, J.A., Rodríguez, S., Alonso, R.S., Corchado, M.: Integrating hardware agents into an enhanced multi-agent architecture for ambient intelligence systems. Inf. Sci. **222**, 47–65 (2013)

10. Tapia, D.I., Alonso, R.S., De la Prieta, F., Zato, C., Rodríguez, S., Corchado, E., Bajo, J., Corchado, J.M.: SYLPH: An Ambient intelligence based platform for integrating heterogeneous wireless sensor networks. In: IEEE International Conference on Fuzzy Systems (FUZZ), pp 1–8 (2010)

11. Anastasopoulos, M., Niebuhr, D., Bartelt, C., Koch, J., Rausch, A.: Towards a reference middleware architecture for ambient intelligence systems. In: Proceedings of the Workshop for Building Software for Pervasive Computing. ACM Conference on Object-Oriented Programming, Systems, Languages, and Applications (2006)

12. Ranganathan, V.K., Siemionow, V., Sahgal, V., Yue, G.H.: Effects of aging on hand function. J. Am. Geriatr. Soc. **49**(11), 1478–1484 (2001)

13. Moreno, A., Nealon, J.L.: Applications of Software Agent Technology in the Health Care Domain. Whitestein Series in Software Agent Technologies. Birkhauser, Switzerland (2003)

14. EUROSTAT, http://epp.eurostat.ec.europa.eu/cache/ITY_OFFPUB/KS-NK-03-026/EN/KS-NK-03-026-EN.PDF (Last visit: November 2012)

15. Insituto Nacional de Estadística, http://www.ine.es/jaxi/menu.do?type=pcaxis&path=/t22/p320&file=inebase&L=0 (Last visited: 13th June 2014)

16. Office for Disability Issues, http://odi.dwp.gov.uk/disability-statistics-and-research/disability-facts-and-figures.php#imp (Last visited: November 2012)

17. Tapia, D.I., De Paz, F., Rodriguez, S., Bajo, J., Corchado, M.: Providing home care using context-aware agents. Int. J. Reasoning-based Intell. Syst. **2**(2), 125–132 (2010)

18. Alonso, R.S., García, O., Zato, C., Gil, Ó., de la Prieta, F.: Intelligent agents and wireless sensor networks: a healthcare telemonitoring system. PAAMS (special sessions and workshops), pp. 429–436 (2010)

19. Corchado, J.M., Bajo, J., Tapia, D.I., Abraham, A.: Using heterogeneous wireless sensor networks in a telemonitoring system for healthcare. IEEE Trans. Inf. Technol. Biomed. **14**(2), 234–240 (2010)

20. Foster, D., McGregor, C., El-Masri, S.: A survey of agent-based intelligent decision support systems to support clinical management and research. In: Proceedings of MAS*BIOMED'05, Utretch, Netherlands (2006)

21. Fiol-Roig, G., Arellano, D., Perales, F.J., Bassa, P., Zanlongo, M.: The intelligent butler: a virtual agent for disabled and elderly people assistance. In: Corchado, J.M., Rodríguez, S., Llinas, J., Molina, J.M. (eds.) International Symposium on Distributed Computing an Artificial Intelligence, vol. 50, pp. 375–384, (2008)

22. Muñoz, C., Arellano, D., Perales, F.J., Fontaned, G.: Perceptual and intelligent domotic system for disabled people. In: Proceedings of the 6th IASTED International Conference on Visualization, Imaging and Image Processing, pp. 70–75 (2006)

23. Velasco, J.R., Marsá Maestre, I., Navarro, A., López, M.A., Vicente, A.J., De La Hoz, E., Paricio, A., Machuca, M.: Location-aware services and interfaces in smart homes using multi-agent systems. In: The 2005 International Conference on Pervasive Systems and Computing (2005)

24. Wooldridge, M., Jennings, J.R.: Agent theories, architectures, and languages: a survey. In: Wooldridge, M., Jennings, J.R. (eds.): Intelligent Agents, pp 1–22, Springer, Berlin (1995)

25. Ferber, J., Gutknecht, O., Michel, F.: From agents to organizations: an organizational view of multi-agent systems. In: Giorgini, P., Muller, J., Odell, J. (eds.) Agent-Oriented Software Engineering VI. LNCS, vol. 2935, pp. 214–230. Springer, Berlin (2003)

26. De la Prieta, F., Pérez-Lancho, B., De Paz, J.F., Bajo, J., Corchado, J.M.: Ovamah: multi-agent-based adaptive virtual organizations. In: 12th International Conference on Information Fusion (FUSION), pp. 990–997 (2009)

27. Zato, C., De Paz, J.F., De Luis, A., Bajo, J., Corchado, J.M.: Model for assigning roles automatically in egovernment virtual organizations. Expert Syst. Appl. **39**(12), 1038–10461 (2012)

28. Zato, C., Villarrubia, G., Sánchez, A., Barri, I., Rubión, E., Fernández, A., Rebate, C., Cabo, J. A., Álamos, T., Sanz, J., Seco, J., Bajo, J., Corchado, J.M.: PANGEA—Platform for Automatic coNstruction of orGanizations of intElligent Agents. In: Proceedings of the DCAI 151, pp 229–239. Springer (2012)

29. PANGEA. http://pangea.usal.es/ (Last visited: 13th June 2014)

30. ZigBee Standards Organization: ZigBee Specification. Document 053474r13. ZigBee Alliance (2006)

31. Tapia, D.I., Alonso, R.S., De Paz, J.F., Corchado, J.M.: Introducing a distributed architecture for heterogeneous wireless sensor networks. Distrib. Comput. Artif. Intell. Bioinf. Soft Comput. Ambient Assist. Living **5518**, 116–123 (2009)

32. Liebermansoftware: Wake on LAN technology. White paper. http://www.liebsoft.com/pdfs/Wake_On_LAN.pdf (Last visited: 13th June 2014)

33. Nedevschi, S., Chandrashekar, J., Liu, J., Nordman, B., Ratnasamy, S., Taft, N.: Skilled in the art of being idle: reducing energy waste in networked systems. In: Proceedings of the 6th USENIX Symposium on Networked Systems Design and Implementation, pp 381–394 (2009)

34. Kohn, L.T., Corrigan, J.M., Donaldson, M.S.: To err is human: building a safer health system. National Academy Press, Washington DC (1999)

35. Camarinha-Matos, L.M., Afsarmanesh, H.: Design of a virtual community infrastructure for elderly care. In: Camarinha-Matos, L.M. (eds.) Proceedings of PRO-VE'02, Sesimbra, Portugal (2002)

36. Villarrubia, G., Bajo, J., De Paz, J.F., Corchado, J.M.: Real time positioning system using different sensors. In: 16th International Conference on Information Fusion (FUSION), pp. 604–609 (2013)

37. De Paz, J.F., Rodríguez, S., Bajo, J., Corchado, J.M.: Mathematical model for dynamic case-based planning. Int. J. Comput. Math. **86**(10–11), 1719–1730 (2009)

38. Tapia, D.I., De Paz, J.F., Rodríguez, S., Bajo, J., Corchado, J.M.: Multi-agent system for security control on industrial environments. Int. Trans. Syst. Sci. Appl. J. 4(3), 222–226 (2008)

39. Davidsson, P.: Multi agent based simulation: beyond social simulation. In: Moss, S., Davidsson, P. (eds.) Multi Agent Based Simulation. LNCS series, vol. 1979, pp. 141–155. Springer, Berlin (2000)

40. Unity 3D Engine. http://unity3d.com/ (Last visited: 13th June 2014)

41. García. E., Rodríguez, S., Martín, B., Zato, C., Pérez, B.: MISIA: middleware infrastructure to simulate intelligent agents. In: Abraham, A., Corchado, J.M., González, S., De Paz, J.F.: International Symposium on Distributed Computing and Artificial intelligence, pp. 107–116. Springer (2011)

42. Foundation for Intelligent Physical Agents. FIPA agent management specification, http://www.fipa.org/specs/fipa00001/SC00001L.html (Lastvisited: 13th June 2014)

43. Hart, P.E., Nilsson, N.J., Raphael, B.: A formal basis for the heuristic determination of minimum cost paths. IEEE Trans. Syst. Sci. Cybern. **4**(2), 100–107 (1968)
44. Rodríguez, S., Pérez-Lancho, B., Bajo, J., Zato, C., Corchado, J.M.: Self-adaptive coordination for organizations of agents in information fusion environments. In: Proceedings of HAIS'10. LNAI, vol. 6077, pp. 444–451. Springer (2010)

Working Garment Integrating Sensor Applications Developed Within the PROeTEX Project for Firefighters

Guy Voirin

Abstract In the frame of the European project PROETEX, CSEM continued to develop applications in the field of smart garment for monitoring vital signs and environment of the wearer. This European integrated project grouped 23 partners from all over Europe. The objectives of the project were to develop smart garments for firefighters and rescuers that will help them and their headquarters to know the location, the capacity, and the encountered risks of emergency personnel on the intervention's field. The smartness was given by the network of integrated sensors and the communication capacities of the developed garments. The inner garments integrated life sign monitoring sensors for heart rate (HR), respiration rate, body temperature, etc. The outer garments were equipped with sensors to determine environmental risks (chemical, temperature), the situation of the rescuer (activity, attitude, localization), and the control system with a communication module. The system was tested in the laboratory and the intervention's field. In the laboratory, it was compared with reference measurement systems and proved its functionality. Then it was tested with firefighters and rescuers in training situations with exercises of real rescue and fire operations. It proved its ability to provide relevant information to the management of emergency personnel during operation. Integration of sensors in garments is an emerging technology that will find applications not only for monitoring people in extreme situations but also for monitoring people during sport training or for monitoring patients and elderly people at home.

Keywords Wearable systems · Smart garment · Protection · Monitoring

G. Voirin (✉)
CSEM Centre Suisse d'Electronique et de Microtechniques SA,
Neuchâtel, Switzerland
e-mail: Guy.Voirin@csem.ch

© Springer International Publishing Switzerland 2015
K. Kinder-Kurlanda and C. Ehrwein Nihan (eds.), *Ubiquitous Computing in the Workplace*, Advances in Intelligent Systems and Computing 333,
DOI 10.1007/978-3-319-13452-9_2

1 Introduction

CSEM has been active in the field of on-body monitoring for more than 10 years. Expertise has been acquired on physiological monitoring and parameters like electrocardiogram (ECG), heart rate (HR), peripheral capillary oxygen saturation (SpO$_2$), and skin temperature (ST). More recently, in the frame of the European project BIOTEX, CSEM acquired expertise on sensing microsystems for parameters in body fluids like sweat or wound exudate [1].

The ambitious strategy for sensors' development and integration in textiles (Fig. 1) was to first develop physiological sensors and biosensors for on-body applications and then to integrate them in wearable garments for seamless monitoring of patients, athletes, or workers exposed to extreme conditions, such as firefighters.

The European project PROeTEX is one pillar of this strategy and deals with smart garments to monitor firefighters and their environment during fire interventions. Smart garments in this context means garments integrating a network of sensors, microprocessors, energy sources, and communication tools.

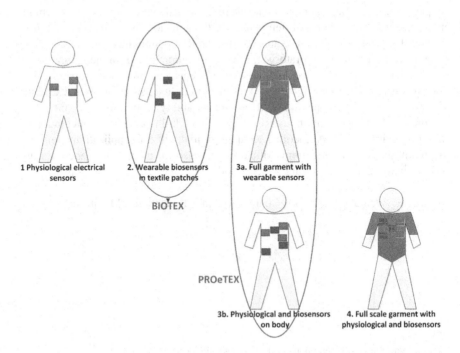

Fig. 1 Strategy for sensor integration in garments (from http://csnej106.csem.ch/sfit/default.htm)

2 General Concept of the PROeTEX Project

2.1 Background

Rescuers and firefighters are professionals that work in difficult conditions. They may be led to put in danger their life for others. In these situations, their judgement can be affected by stress due to physical efforts or temperature. Therefore, it is key for the operation headquarters to know and evaluate the situation in which fire-fighters are engaged, in order to take decisions about the operation's execution. There is a need to determine with accuracy the location, the environmental conditions, and the professionals' ability on the frontline.

In light of the latest developments in wireless communication and the interconnections between the different systems, it has been possible to envisage a development of a system (Fig. 2) with the following capabilities [2]:

- Provide information about the personnel involved on the intervention's field and about their activities.
- Ensure that the entire personnel are monitored to determine whether they become incapacitated and, if this occurs, rapidly set up an extraction plan from the operation's field.
- Use physiological and biological sensors to predict the capacity of engaged personnel to operate by monitoring their level of physical and psychological stress.
- Determine the environmental, thermal, chemical (toxic gases, chemical agents, corrosive vapors) and physical risks (falls, crushes, blasts).

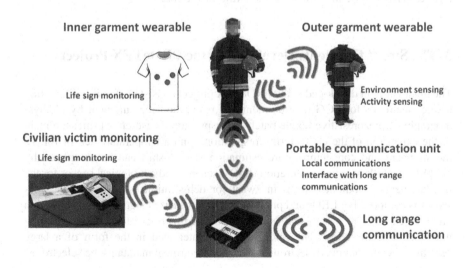

Fig. 2 PROeTEX concept

- Store an individual event log per user for preventive medicine and epidemiological studies.
- Monitor the state of health and location of injured civilians in order to maintain an overall plan of evacuation based on constantly updated relevant information.

These capabilities are particularly important in major, large-scale events such as forest fires, earthquakes, or major terrorist incidents, where large numbers of emergency personnel are working for extended periods of time over large areas and where significant numbers of injured civilians are involved.

2.2 Concept

In order to bring a solution to the rescuers' and firefighters' problems, a European consortium was set up. It comprised more than 20 groups from universities, research centers and companies. The basic concept put forward in the project is to have an inner garment directly on the body with sensors for life signs monitoring, an outer garment with sensors for activity, location, and environmental sensing, and a communication unit that will perform local on-body communications and make the interface with a system with a longer range to be able to connect with the system in use during the operation. The solution also offers a system for life signs monitoring of victims to help rescuers to classify the victims depending on the seriousness of their injuries and the risk for their life and to be able to intervene in case their conditions worsen. One part of the project was dedicated to long-term R&D for the development of fiber-based micro- and nanosystems to prepare the miniaturised sensors to be integrated in the textile of the future.

3 The Smart Garments Developed in the PROeTEX Project

The inner garment developed in the PROeTEX project consisted of a shirt integrating textile electrodes for ECG measurement and respiration measurement by plethysmography, a piezoresistive textile patch for respiration measurement (measurement of the expansion of the chest during respiration), and a temperature sensor for ST measurement and core temperature estimation. The T-shirt can be completed by peripheral capillary oxygen saturation (SpO_2) sensors and a biological sensor for the measurement of sodium content in sweat for dehydration estimation. The SpO_2 sensor is composed of LEDs and photodiodes. The different signals are processed in order to extract the best signal and to reduce artifacts created by the different movements of the body. The SpO_2 sensor is integrated in the form of a large "electrode" with integrated electronics. The elastic garment maintains the "electrode" in contact with the skin. It is an alternative to electrodes and textile being directly incorporated into the fabric.

The outer garment is composed of the main electronic system, which is the base of the body network. This network links different sensors together: a GPS module for position determination in outdoor situations, several three axis accelerometers for activity and posture determination, the sensors for monitoring the environment of the firefighters including a temperature sensor, a heat flux sensor, and a gas sensor. This wired sensor network is extended to sensors in the boot using a ZigBee-like wireless module. The network is linked with a Bluetooth communication module to the long-distance communication module (Wi-Fi) via a textile antenna. The network also includes alarms that can be activated by the system to alert the firefighters when a specific dangerous situation is detected. The alarms are acoustic and optical, a buzzer and high-power flashing LED lights, respectively. The system also has energy sources; two types were used: a flexible prototype developed at CEA (France) for direct integration into the garment and a commercial Li-ion battery pack (Fig. 3).

The electronic system was integrated into the garment in different pockets and by guide wires sewn into garments. Several releases of the system were manufactured and tested.

4 The Tests of the Firefighter Garments

Two different types of tests were made with the system. Laboratory testing was conducted to verify the functionality of the system and to verify the accuracy of the different sensors using gold standard devices for comparison.

The ergonomics of the system were tested with the firefighter and the rescuer team. The goal of these tests was to show that the system would not be a disruptive factor due to the size of some of the electronic devices. The system was easily accepted by the firefighters because their usual outer garment is already quite heavy and the use of electronic devices does not make the situation worse. Moreover, during the interventions, firefighters often wear respiratory equipment, which is by far more cumbersome.

After the laboratory tests, the system was tested and demonstrated in exercises simulating real situations. The exercises were done in a test location for firefighters and rescuers in Italy. Different exercises were performed by firefighters and rescuers like extinction of fire, carrying victims, or smoke divers' training. Figure 4 presents several pictures taken during these exercises.

The signals of the different sensors were recorded using the system, sent using the wireless communication element, and displayed on the computer representing the headquarters of the operations (Fig. 5).

(a)

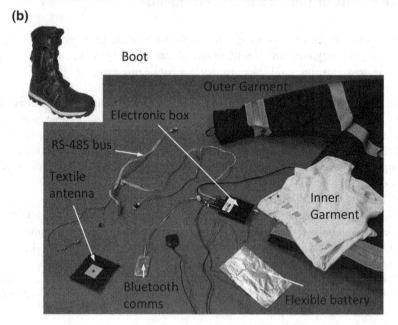

(b)

Fig. 3 **a** Block diagram of the electronic system. **b** View of a first release of the garment prototype with the sensor network

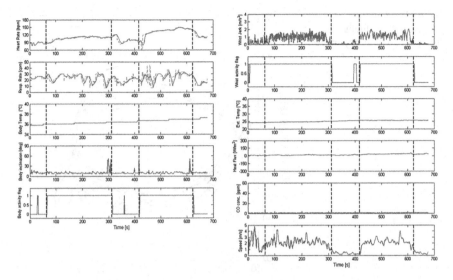

Fig. 4 Pictures of firefighters during exercises simulating operation in the field

Fig. 5 Display of the signals of the different sensors on the computer representing the headquarters of the operation

The location of firefighters or rescuers is done using GPS; this works well outdoors, but it is not adapted to firefighter operations inside buildings. The position is acquired with the wearable electronic device and sent to the supervising computer where the data can be displayed either as x, y coordinate (Fig. 6) or superimposed on a map or satellite picture in a localization software.

Fig. 6 Trace of the x, y position of three firefighters during an exercise of portage

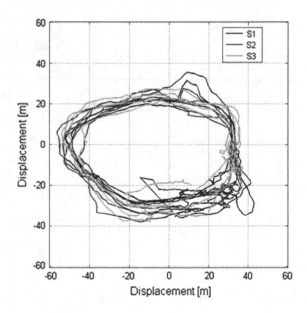

The system's features are summarized in the following table:

- Inner garment:

    - HR (ECG): 50–240 bpm
    - Respiration rate: 5–60 bpm
    - Skin temperature: at 30 °C, accuracy ±0.33 °C; in the range 10–50 °C, ±0.5 °C
    - Na^+ concentration in sweat
    - SpO_2: 70–100 %, resolution 1 %, accuracy ±3 %

- Outer garment:

    - Localization: GPS
    - External temperature: −70 °C, 500 °C; heat flux
    - Two 3 axis accelerometers for movement and posture determination
    - Environmental gas sensing: CO, CO_2
    - Alarms: visual, acoustic

5 Outlook and Conclusion

Within the PROeTEX project, new ways of integrating sensors in textiles were tested. Different sensors with fiber geometry were tested like organic transistors, organic LED, and biosensing optical fibers. In view of the integration of wearable sensors into textiles, the fiber geometry offers an attractive support for sensors as they can be woven like standard yarn, to some extent, and offer minimally invasive, on-body, highly sensitive measurements of health monitoring. At CSEM, biosensing optical fibers were developed; they consist of optical fibers with a sensing layer comprising the chemical or biological sensing element, a light source, and a detector (photodetector, spectrometer). Multiple reflections of the light propagating in the optical fiber allow sensing of optical changes in the proximity of the fiber core—within the evanescent field. Therefore, the sensitive layer is placed directly on the core of the optical fiber, replacing the original cladding that confines light in the core of commercial glass or polymer fibers. Sensing optical fibers have been designed for performing pH measurements of sweat by incorporating a pH-sensitive dye in the sensitive layer deposited on the fiber core. For the detection of bioanalytes such as lactate or glucose, work has started for the development of sensitive layers incorporating specific enzymes (e.g., lactate or glucose oxidase). These sensitive fibers are promising for the sensing of biological parameters on body as they will give more information about stress, tiredness, or dehydration of the firefighters.

In the PROeTEX project, garments were developed to help workers to have a better knowledge of their environment and their capacities and to receive information through simple acoustic or optical alarms. The next step for "smart garments" will be to have a direct action in the worker environment, for example, in the project prosys laser [3], where development was dedicated to the protection of workers using a handheld high-power laser processing system, by using active functional multilayer textiles incorporating sensors that detect laser exposure. By means of a safety control unit (the smartness of the garment), the protective cloth is able to deactivate the laser beam automatically for the safety of the worker.

The PROeTEX project has shown that it is possible to develop garments that integrate numerous sensors that can monitor parameters of the environment and of the biophysiological state of the workers in extreme conditions. In the future, the "smart garment" will have more and more interaction with the users and will even modify their environment for their safety. PROeTEX has also demonstrated that the development of wearable sensors can follow two ways, either as an electronic component integrated into textiles or as sensitive fibers directly integrated into the textile fabrics. The PROeTEX developments are the premises of applications that will not be reserved to firefighters and rescuers but will also find applications in the field of sport training and medical devices.

Acknowledgments The authors want to thank all the partners of the PROeTEX project. This work was partly funded by the European Commission project FP6-IST-026987.

References

1. Pasche, S., Angeloni, S., Ischer, R., Liley, M., Luprano, J., Voirin, G.: Wearable biosensors for monitoring wound healing. Adv. Sci. Technol. **57,** 80–87 (2008)
2. http://www.proetex.org
3. http://www.prosyslaser.eu/

Part II
Cross-disciplinary Perspective on Ethical Issues

The Question of Ethics in Ambient Intelligence

Klaus Wiegerling

Abstract Fundamental ethical issues regarding ambient intelligence concern the conditions of an ethical discourse, the identity of the acting individual, the awareness of the sphere in which the individual should act, and finally the individual's choice, which is necessary for his or her responsibility. Augmented reality exacerbates the problem of the loss of resistance and horizon which characterize what the German term "Wirklichkeit" means. Only the hyletic, social, or ideal resistance against our will to shape the world and the horizon of connected systems guarantees "Wirklichkeit," but not the real, concrete given thing, because there is always a possibility of misjudging it in perception. We are confronted with a loss of the horizon, in which we have to act. The old paradigm of an embedded technology is coming to a critical point now, if smart systems shape the world for us without a possibility to control and guide this shaping. Autonomous systems have a capacity to incapacitate the user.

Keywords Reality · Wirklichkeit · Resistance · Horizon · Self-determination · Relief · Incapacitation

1 Introduction

It is not my intention to speak of the positive aspects of new visions in computer science. Besides mediation and orientation, the most important task of philosophy is a negative one, namely to criticize current ideas and visions, even though hopes for a better life are connected to doing so. Neither will I give an introduction to these visions with attempts to define special terms or to ascertain different concepts. First, I think the reader is familiar with these visions and concepts theoretically and probably with some implementations practically. I will expose only some essential

K. Wiegerling (✉)
Fachgebiet Philosophie, Universität Kaiserslautern, Kaiserslautern, Deustchland
e-mail: wiegerlingklaus@aol.com

© Springer International Publishing Switzerland 2015
K. Kinder-Kurlanda and C. Ehrwein Nihan (eds.), *Ubiquitous Computing in the Workplace*, Advances in Intelligent Systems and Computing 333,
DOI 10.1007/978-3-319-13452-9_3

characteristics of the most important ideas in order to understand the respective ethical discourse. Second, it is a problem to use technological terms and concepts in a philosophical way, because these terms and concepts are continuously changing according to the preferences of society and technological proficiencies and experiences. New technological and judicial problems appear with the development of the technology we are talking about. There is both an extension and a contraction of goals, because technology is always part of the progress or change of the society into which it is embedded.

2 The Basic Questions of the Ethical Discourse

All ethical discourses contain three basic questions which in themselves are strictly speaking not ethical by nature. These questions are meta-ethical ones, because indeed they have an epistemological or logical rather than a normative meaning. But they are fundamental for an ethical discourse in general. The first question concerns the identity of the acting subject who is to be responsible for his or her actions. If this identity is weakened, such as in the case of a schizophrenic, the individual is not necessarily accountable for his or her own actions. The second question is that of the "Wirklichkeit" in which we have to act. In German, there are two different concepts for the English term "reality": that of "Realität" and that of "Wirklichkeit." For now, I will provide only some general remarks on these terms. If we are to actively operate in the world, we have to be able to define "Wirklichkeit" in the sense of a "lived-in world" or lifeworld, i.e., not in an absolute, but in a relative or practical sense. We need a definition that complies with the demands of everyday life. For example, we have to know that it is not deemed proper to interfere when Wallenstein is murdered on stage. Finally, the third question is that of choice. We can be responsible only for an action which we have chosen. Thus, choice is the third fundamental condition for an ethical discourse.

In light of ever smarter new and autonomously operating information systems, these three basic questions profoundly touch on the ethics of media particularly, but also on ethics in general. The experience of "Wirklichkeit," the development of a personal identity, and the possibility to choose will be different in intelligent environments modified by ubiquitous systems, because resistance, the confrontational character which is as vital for the experience of "Wirklichkeit" as it is for the development of personal identity, has been at least partially removed by the nearly complete loss of the interface.

3 The Vision of Ambient Intelligence

Let us have a look to the visions of ambient intelligence, ubiquitous computing, and pervasive computing. The technologies, which are described by these terms and of which different characteristics are emphasized by each, offer a vision which can be realized in different technologies. Thus, these terms signify a vision, not a concrete technology.

Compared to pervasive computing, the term ambient intelligence is used to emphasize more strongly the social embeddedness of the technology and the interaction with the system. Ambient intelligence denotes a kind of intelligence which can be found in our immediate environment in the sense that this intelligence, which is also a vital part of the lifeworld term, is readily available to us around the clock. That means it is part of the lifeworld, of the "Lebenswelt." The concept of "Lebenswelt" was originally developed by Edmund Husserl as a specifically philosophical term. According to Husserl, the Lebenswelt as a world which we do not question underlies all types of worlds in the positive sciences. All scientific types of worlds result out of a reduction of the Lebenswelt [1].

The focus of ambient intelligence is (1) the context of information, which means the social and the physical context, (2) the confidential treatment of personal data, (3) the use of other user interfaces than touch screen or mouse, for example, physical elements such as writing utensils, and (4) the personalized use of the system according to a person's preferences. However, the three terms of ubiquitous computing, pervasive computing, and ambient intelligence only focus on different aspects of the same technological vision, which was introduced by Weiser [2], but not really on different technologies [3].

We can point out the following characteristics of this vision of informatics: the general disappearance of man–system interfaces and hardware components; the adaptiveness, smartness, context awareness, and self-organization of the system; the augmentation of reality by the system; the ubiquitous, anytime use of the system as well as the linking of global and local information relating to the situation in which we have to act.

4 The Basic Problem of Smart Systems: The Loss of Obtrusiveness

For a long time, the criterion for a well-introduced technology was the loss of obtrusiveness. Like perfectly fitting glasses are so to speak a quasi-"natural" part of us, technology should become seamlessly integrated into our everyday life and should be unobtrusive. Thus, smart systems should be adapted in the same way as perfectly fitting glasses are. But the problem of the "traditional" criterion for a well-introduced technology in the case of smart systems is that we get into a new dependency, because the functioning effects of smart systems become invisible and because the obtrusiveness or resistance of things is no longer perceivable. We are being confronted only with things that are already in an augmented mode, but augmented often only in certain aspects, for example, economic ones. But we can only control something and thus take responsibility for an effect, if we are aware of it.

Indeed, this is a critical view of a fundamental concern of ambient intelligence and may be a challenge for the vision, but I think it is a necessary challenge if we want to avoid the problem which Goethe has pointed out in his famous poem about the sorcerer's apprentice.

On the other hand, the concept of reality, "Wirklichkeit," which is, according to Dilthey [4], characterized by a resistance against one's will to form or to change it and by symbolic links to concrete parts of reality, is gradually altered into an optional state, which is not defined by concrete distances and concrete relations. This is done by systems that augment the always concrete, given pieces of reality without one actually having to operate them. Reality is not actuality and "Wirklichkeit" is always more than actuality, because it transcends it. "Wirklichkeit" is always given in a historical or symbolic kind, and its agency is not only a current one.

"Actuality" can be misunderstood in a sense of an effect, which is given now in this moment. But historical effects are not always given in a sense of current awareness; normally, we are not conscious of these effects.

The effects of the system are often given with delay and we cannot relate an effect to its cause. However, the resistance "against" us characterizes exactly that, what "Wirklichkeit" means. If a system becomes smart, this characterizing point is expressed in a virtualization.

Furthermore, "Wirklichkeit" is a specific connection of different pieces of reality. Thus, the "Wirklichkeit" of a person living in the Australian bush is completely different from the "Wirklichkeit" of a person living in Central Europe. Thus, "Wirklichkeit" unifies two aspects: the resistance to one's will to form or to change it and the connection of different pieces of the concretely given reality.

It is not possible to experience "Wirklichkeit" in a concrete way, but it is also not possible to experience reality beyond a concept of "Wirklichkeit." "Wirklichkeit" is the specific horizon, in front of which we can catch reality as a concrete given thing. That means our concept of the given thing is determined by this horizon. "Wirklichkeit" is not a creation of one's own, but rather a creation of the culture which we are part of.

We do not usually experience the world like the artist or the engineer do, who view it as a potential which can be molded and shaped in various ways. Rather, we see the world as something that has been created for us to be used in a certain way —namely as a typical user. We are not the creators of "Wirklichkeit," but the system is. The system is, so to speak, an expression of a specific culture, but only one and only a very particular one. It is not an expression of culture in general. It gives a connection of the concrete pieces of reality in a way that is assumed to be helpful for us in a specific situation. Thus, the system creates "Wirklichkeit" by integrating given pieces of reality into an imagined "Wirklichkeit," but removes the aspect of resistance. Therefore, "Augmented Reality" is actually an expression of Wirklichkeit, because it provides us with a connection between concrete pieces of reality.

Man–system interaction in ambient intelligence is reduced to a minimum, as it only occurs to provide the initial spark. There are more reactions of the system without any order from the user. The problem is that the system does not only support the single individual. The system has a capacity to substitute any single person's abilities by virtual agents and to guide their wishes in a certain direction, driven by political or economic interests.

The basic problem in augmenting reality through information systems is the possible loss of perception of resistance. If the environment is becoming intelligent and if reality is augmented by information, one's intentions are regulated by autonomous systems and the elementary characteristics of "Wirklichkeit," the resistance against one's will to shape things is no longer present. As it is self-organizing, the system becomes, so to speak, the acting "person." Although tasks are delegated to the system by individuals, they are carried out in a way, which is neither delegated nor wished by its nominal users. The user is treated as a stereotypical user, as part of an anonymous group. It is not possible to cancel the membership to this anonymous group because users are not aware that they are members of it. Thus, the information system operates in a way which cannot be individually controlled.

According to paternalistic effects, the systems used are an expression of the society and its preferences and claims, so to speak a machine that moderates individual wishes and intentions in a common sense.

The context sensibility of a system is based on a program of decontextualization, i.e., it works by reducing services, situations, and behaviors to match typical stored data bits or by disarticulating certain areas of "Wirklichkeit" and thus exacerbates the mentioned problem. Faulty linking is likely when systems can no longer be actually operated and controlled, i.e., when the option to interact with the system is no longer possible because the interface of the system is no longer perceivable.

When hierarchies in systems and the borders of systems have become invisible, the result will be less individual competence, and thus, the development of personal identity will be affected, because every personal identity is formed by acknowledgment and non-acknowledgment of his or her actions and by the development of individual competence. That means the development of routines and techniques by which intentions can be realized and the resistance of the world can be arranged in a bearable way. Since systems exchange, process, and trade autonomously acquired data, the process of how the system arrives at a certain conclusion can no longer be easily traced. Thus, we are faced with the challenge of how to preserve the resistance inherent in "Wirklichkeit" as a basic experience necessary for any cultural and explicit technical activity. In fact, we can say that the experience of resistance and the formation of resistance is an elementary expression of life.

But the basic problem of the loss of resistance is not to understand it in a wrong sense. It does not mean that individuals must develop their personal identities in a permanent struggle against the obstacles and contingencies of everyday life. This would question any means to make life easier, which is the condition for many higher cultural achievements. The reduction of resistance finally is the sine qua non condition for any cultural progress or development, respectively. But if the means to make our life easier are not perceivable, and if they disappear completely and merge with the material mesosphere, then we have no chance to control the system's functions and effects. Thus, it is necessary to be aware of these functions and effects; otherwise, we do not act autonomously. If we cannot notice how the system works and what the system affects, it slips out of our hands and degenerates into an apparatus for incapacitation. The system will tell us what we have to do, either in a direct way or in an indirect way. Thus, the focus in developing information systems should be on making perceivable possibilities to control autonomously operating systems. Without doubt, in the visions of ambient intelligence, there are capacities for incapacitation of the user or the beneficiary, of this technology.

5 The Anthropological Dimension of the Loss of Resistance

The problem of the loss of the resistance that characterizes "Wirklichkeit" gets an anthropological dimension by the enhancement of the human body with implants and prostheses which are connected informatorily with an ambient intelligence. In the idea of transhumanism for the future human being—maybe a super-human like in Nietzsche's concept of the "Übermensch" and maybe a completely new species in which the ideas of cyborg and biofact converge—the skin is not the external border of the human body and this body is integrated into an all-encompassing intelligent environment. The confrontational character of the "Wirklichkeit" then is possibly or even probably filtered by systems that operate intracorporally. We are aware of what we should be aware of, and it is quasi-dictated by the reason of the system. The idea of enhancement, not least by intelligent implants and prostheses which are connected with extracorporal systems, opens a wide sphere to discuss the anthropological dimension of ambient intelligence [5]. This concerns our idea of humanity and our relation to the world in a fundamental sense. We have to ask what humanism means. Finally, the term humanism is a cultural–historical and specifically ethical category, which cannot be used independently of its historical state. But what does this state mean in a technological conception? Maybe we are on the way to substituting not only the physiological disposition of man, but also the historical and cultural disposition of the human race in general, because the future being which is following the human being which we know so far will not have a history, either an individual or a collective one, in the current sense. The future being at the most is part of a technical development, which means that it is not really an individual with an individual history. Maybe this future being will have no interest in negotiation, but instead in an efficient realization of his or her wishes, which are completely compatible with the community and the system which represent this community.

6 Exemplification: The Dialectic of Relief and Incapacitation by the Application of AAL Systems

Let us look to some current applications that are based on the idea of ubiquitous computing to show how the meta-ethical condition of choice is concerned.

In care of the elderly or health care, it is hoped that with the help of ubiquitous systems, called ambient assisted living (AAL) systems, it will become possible to handle the problems of demographic change. Elderly people should be enabled to live autonomously in their own home for longer with the help of AAL systems. The systems, for example, could remind elderly people to take their medication, could manage their everyday lives, could monitor their state of health, and could assist them with robotic systems. Lastly, the systems make sure that the user is permanently connected with their relatives, friends, and all social institutions that are necessary for being comfortable and happy. But naturally, the important function of relief provided by AAL systems can change into incapacitation. The institution of assistance systems in health care and care of the elderly is a balancing act between relief and incapacitation. Paternalistic effects probably relieve society at the cost of the individual's majority. Also, there is no need for personal assistance if a system can help the individual. But an elderly, diseased, or disabled person is not necessarily unable to decide on his or her own affairs. Thus, there is a conflict between personal will and freedom on the one hand and reasonable constraint through the AAL systems on the other hand, especially if the systems do not allow one to escape from assistance. An individual has no majority and is free from responsibility, when he/she has no choice. It is difficult to say whether paternalistic effects from AAL systems are avoidable at all. Certainly, such effects are pursued by assistance systems, if alternatives of acting or possibilities to escape from assistance are not announced. Within the concept of persuasive computing, these effects even achieve a strategic status.

Here, we get to a point of misunderstanding with regard to this concern. Naturally, it is helpful if an elderly individual who has difficulties with orientation or a person who is mentally ill is permanently monitored by such a system. But not all old, disabled, or mentally fragile people are unable to decide their own affairs. It is furthermore an expression of human dignity and freedom that all people can decide against reason or what the majority of the society believes that reason is. But without an alternative to the system's assistance for such individuals, there is no choice, which means no freedom, no responsibility, and indeed no personality in a strict sense.

7 Résumé

Let us summarize the central points of this paper: Fundamental ethical issues regarding new technological visions in informatics concern the conditions of an ethical discourse, the identity of the acting individual, the awareness of the sphere in which the individual should act, and finally the individual's choice, which is necessary for responsibility. An augmented reality exacerbates the problem of the loss of resistance, which characterizes "Wirklichkeit."

There is a conflict between using AAL systems and the principle of subsidiarity which is a protection against the incapacitation of man. That means the use of these systems has a specifically political dimension. The question is whether we are ready to give up the fundamentals of a democratic constitution: the autonomy of the individual and its dignity which is articulated by this autonomy.

The old paradigm of embedded technology is coming to a critical point now, as smart systems shape the world for us without a possibility to control and guide this shaping.

It is our responsibility to give an answer to the question of whether the vision of smart ubiquitous systems that accompany us in our everyday lives should be challenged totally or only in some parts. I believe that it is a challenge only in some parts, namely those that concern the configuration of the systems. But the idea of smartness and autonomy of the systems by virtual agents requires new discussions. The problem is not that technology works in the background—many technologies do that—the problem is that smart and quasi-autonomous technologies also have the capacity to incapacitate the user.

References

1. Husserl, E.: Die Lebenswelt—Auslegungen der vorgegebenen Welt und ihre Konstitution. Texte aus dem Nachlass (1916–1937), Springer, Dordrecht (2008)
2. Weiser, M.: The computer of the 21st century. Sci. Am. **265**(3), 94–104 (1991)
3. Wiegerling, K.: Philosophie Intelligenter Welten, pp. 19–25. Wilhelm Fink, München (2011)
4. Dilthey, W.: Beiträge zur Lösung der Frage vom Ursprung unseres Glaubens an die Realität der Außenwelt und seinem Recht (1890). In: Dilthey, W.: Gesammelte Werke 5. Die Geistige Welt – Einleitung in die Philosophie des Lebens, 8th edn., pp. 90–138. Vandenhoeck & Ruprecht, Göttingen (1990)
5. Wiegerling, K.: Zum Wandel des Verhältnisses von Leib und Lebenswelt in Intelligenten Umgebungen. In: Fischer, P., Luckner, A., Ramming, U. (eds.) Reflexion des Möglichen – Zu Christoph Hubigs Philosophie der Medialität, pp. 225–238. Lit, Münster/Westf. (2012)

Ethical Issues in Ubiquitous Computing—Three Technology Assessment Studies Revisited

Lorenz M. Hilty

Abstract This paper discusses ethical issues in ubiquitous (or pervasive) computing from the perspective of the general discourse on ethics in computing, which started in the 1970s, two decades before the "ubicomp" vision emerged. The IFIP "Human Choice and Computers" (HCC) conferences are used as points of reference for the general computing ethics discourse, and three technology assessment projects related to the ubicomp vision serve as a (nonrepresentative) sample of documents from the discussion of ethical issues in a ubicomp world. Revisiting these studies from the general computing ethics point of view shows that the basic issues have persisted, but ubicomp has added new aspects that were not anticipated in the earlier discourse.

Keywords Ubiquitous computing · Ethics · Autonomy · Responsibility · Sustainability · Justice

1 Introduction

The terms "ubiquitous computing" (or "ubicomp" for short), "pervasive computing," "ambient intelligence," and "the Internet of Things" refer to technological visions that share one basic idea: to make computing resources available anytime and anywhere, freeing the user from the constraint of interacting with ICT devices

L.M. Hilty (✉)
Department of Informatics, University of Zürich, Zurich, Switzerland
e-mail: hilty@ifi.uzh.ch

L.M. Hilty
Technology and Society Laboratory, Empa, Swiss Federal Laboratories for Materials Science and Technology, St. Gallen, Switzerland

L.M. Hilty
Centre for Sustainable Communications (CESC), KTH Royal Institute of Technology, Stockholm, Sweden

© Springer International Publishing Switzerland 2015
K. Kinder-Kurlanda and C. Ehrwein Nihan (eds.), *Ubiquitous Computing in the Workplace*, Advances in Intelligent Systems and Computing 333,
DOI 10.1007/978-3-319-13452-9_4

explicitly via keyboards and screens. This is possible by invisibly embedding computational devices in everyday objects and equipping them with sensors that enable them to collect data without the user's active intervention or even awareness.

This vision has partly become a reality during the last two decades through the continued miniaturization of ICT devices, the use of positioning systems making devices aware of their location, and the growth of networks for wireless or mobile communication. Ubiquity of ICT can even be understood at a global scale today, given the success and impact of the mobile phone particularly in the poor and heavily populated regions of the globe. However, some aspects of the ubicomp vision have not been realized (yet)—for example, we are still using screens to interact with smartphones and many other ICT devices. Conversely, technologies have emerged that had not been anticipated in the ubicomp vision, such as the availability of drones carrying cameras and wireless communication devices that are even affordable for private users.

This essay aims to identify the main ethical issues emerging from the vision and practice of ubiquitous computing. If we assume that an "applied ethics of *ubiquitous* computing" is different from an "applied ethics of computing," there must be ethical issues specifically connected to the ubicomp vision and practice. Hence, the precise question I am trying to answer in this essay is, "What are the *specific* ethical issues in ubiquitous computing, viewed against the background of the (general) ethics of computing?"

The method for answering this question consists of three steps:

1. Identifying the main ethical issues that have been discussed in ethics of computing since the discourse emerged in the 1970s. This will be done by taking the discourse documented in IFIP proceedings as a reference.
2. Identifying ethical issues emerging from the ubicomp discourse, which emerged around the year 2000. This will be done by evaluating three technology assessment studies related to ubiquitous computing.
3. Classifying these issues either as special cases of preexisting more general issues or as new issues which have not been discussed before.

The scope of this work will be limited by focusing on three technology assessment studies from which the ubicomp ethical issues are derived. The sequence of these three studies, selected from the studies published by the Swiss Centre for Technology Assessment (TA-SWISS), starts with possibly the first technology assessment study on ubiquitous computing ever conducted (the project started in 2002) and ends with one of the most recent ones (published in 2012). Taking this sequence as *pars pro toto* for the development of the discourse on implications of ubiquitous computing is obviously a limitation of the current analysis. However, any wider-ranging approach would go beyond the scope of this short essay.

2 Materials and Method

Historically, the discourse on ethics of computing has been initiated and constantly promoted at the international level by IFIP TC9, IFIP's Technical Committee on ICT and Society. IFIP, the International Federation for Information Processing, was founded in 1960 under the auspices of UNESCO as an umbrella organization of the national computer societies. IFIP TC9 has continuously inspired, monitored, and framed the development of the national ethics guidelines and codes of conduct for computer professionals in the national member societies [3].

The work of IFIP TC9 can therefore be used as a reference for the development of the ethical discourse in computing. Instead of digging into the historical details of the development of ethics codes and guidelines, the following analysis will rather take a "helicopter view" and look at the broader discourse documented in the proceedings of the "Human Choice and Computers (HCC)" conference series, IFIP TC9's main conference. The analysis will rely on a recent lexicometric discourse analysis of the HCC proceedings from 1974 to 2012 [2, 4–6, 8, 11, 25, 26, 28, 29] conducted by Lignovskaya [24]. By providing the wider context in which ethical issues in computing have emerged over four decades, the HCC proceedings are an invaluable source of understanding of today's ethical concerns in computing.

There is an important structural difference between the general computing discourse and the ubicomp discourse: While the former emerged in the 1970s when computers had already begun to change everyday reality (in particular in the workplace), the ubicomp discourse started *before* ubicomp became reality. Even today, essential aspects of ubicomp are far from common. Ethical issues of ubicomp are therefore, at least in part, associated with *prospective* applications of computing, not necessarily only with applications existing today.

The public discourse on potential positive and negative impacts of prospective technological applications is often initiated and driven by institutions of *Technology Assessment (TA)*. TA is the study and evaluation of new technologies that are relevant for society and have ethical implications. Probably, the first TA study on ubicomp (in that case called "pervasive computing") was commissioned in 2002 and published in 2003 by TA-SWISS. An English translation of the 354-page study was published jointly by TA-SWISS and the Scientific Technology Options Assessment (STOA) body at the European Parliament in 2005 [13]. Since then, TA-SWISS has commissioned and published two additional studies related to ubicomp, one broaching the issue of the increasing autonomy or emancipation of computers [9], published in 2008, and a recent study on technologies for locating, tracking, and tracing [18], published in 2012.

The reason for selecting these three studies is that they emerged in a uniform institutional context (TA-SWISS), spanning a decade from the first systematic approach to assessing the implications of ubicomp to the most recent study. A review of the entire body of TA studies related to ubicomp would certainly provide a more comprehensive picture, but also go beyond the scope of this essay. Besides this geographic and institutional bias, this paper may also have a personal bias because the author has been involved in two of these studies. I hope that the reader will nevertheless benefit from the—partially subjective—perspective presented in this paper.

The materials used for this analysis are therefore:

1. As a reference for the general discourse on ethical issues in computing: The "HCC" proceedings published by IFIP in the period 1974–2012 [2, 4–6, 8, 11, 25, 26, 28, 29] and, as a secondary source, the discourse analysis conducted by Lignovskaya [24] on these proceedings.
2. As sources for identifying ethical issues in ubiquitous computing, the following are three TA-SWISS studies and related literature:

   (a) TA 46e/2005: "The Precautionary Principle in the Information Society: Effects of Pervasive Computing on Health and Environment" [13] and the related articles [12, 16, 30, 31][1];
   (b) TA 51/2008: "Die Verselbständigung des Computers" ("The Emancipation of the Computer"), published in German [9]; this study covers an essential implication of the ubicomp vision, the increasing autonomy of computers;
   (c) TA 57/2011: "Lokalisiert und identifiziert. Wie Ortungstechnologien unser Leben verändern" ("Located and Identified. How Positioning Technologies Are Changing Our Lives"), published in German [18], and an international conference paper summarizing the study [19]; this study focuses on one essential aspect of ubiquitous computing, the increasing location awareness of objects.

Besides these main sources, additional literature will be used where appropriate to illustrate or support the argument. In particular the work of the "Ad Hoc Committee for Responsible Computing," an international group that developed a "normative guide for people who design, develop, deploy, evaluate or use computing artifacts" [1] will be considered as an additional input on applied ethics of computing, as well as the report "Exploring the Business and Social Impacts of Pervasive Computing" [20], jointly edited by IBM Research, the reinsurance company Swiss Re, and TA-SWISS, on specific ubicomp issues.

I will first identify the invariants in the discourse documented in the HCC proceedings in order to reveal the ethical issues of computing that seem to persist over time (although with a change in focus). In the second step, I will analyze the three TA studies, identifying ethical issues emerging from the ubicomp discourse.

3 Results

The persistent themes in the discourse on ethics of computing as documented in the HCC proceedings from 1974 to 2012 can be subsumed under three umbrella themes:

---

[1] "Pervasive computing" is considered synonymous to "ubiquitous computing" in this context.

- Autonomy and self-determination
- Responsibility
- Distributive justice.

The definitions of the umbrella themes are provided in the following subsections. This classification is not intended as a conceptual framework, but as a pragmatic means of structuring the issues found in the discourse analysis. The umbrella themes overlap, and some ethical issues may therefore be subsumed under more than one of them.

One result of this study is that all major issues discussed in the three ubicomp studies can be matched with the preexisting ethical issues (as shown in Tables 1, 2, 3), however, with new aspects occurring at a more concrete level.

Table 1 Results for autonomy and self-determination

Ethical issues in computing	Ethical issues in ubiquitous computing
Working conditions: • effects of computerization on job satisfaction [4, 25, 26, 29] • participation [4]	Working conditions: • surveillance of employees [18] • blurring boundaries between private and professional life [18]
Virtual and augmented reality: • avatars [5, 28] • virtual property [6]	Virtual and augmented reality: • in remote diagnosis [13] • in surgery [13]
Privacy: • informational self-determination [11, 28, 29] • and biometrics [8] • in health care [5] • and social media [11]	Privacy: • automatic identification [13, 18] • location privacy [18] • implications of transparency [9]
Technology paternalism: • security and biometrics [8, 11] • in e-health [5]	Technology paternalism: • as a tendency in ubicomp [9] • by the use of active implants [13] • in dependency relationships [18]

Table 2 Results for responsibility

Ethical issues in computing	Ethical issues in ubiquitous computing
Legal and moral responsibility • for "computer decisions" [8, 25] • in e-health [5] • of the user of social media [11]	Legal and moral responsibility • and autonomous computer systems [9] • the "dissipation" of responsibility [13]
Social responsibility • for the impacts of automation and globalization [26, 29] • of governments [29] • of computer professionals [2, 4–6, 8, 11, 25, 28, 29]	Social responsibility • for the implications of transparency [9]

Table 3 Results for distributive justice

Ethical issues in computing	Ethical issues in ubiquitous computing
Digital divide • computer literacy [4, 8, 29] • technology transfer [4] • intellectual property, piracy [6]	Digital divide • reducing the digital divide [13]
Sustainable development • and the information society [2, 5, 6, 8] • sustainable X [11]	Sustainable development • dematerialization potential [13] • material dissipation [13] • creation of a critical infrastructure [18]

3.1 Autonomy and Self-determination

Autonomy, as a philosophical concept, is the capacity of individuals to make choices based on their own personal beliefs and values. If seen as an ethical value, autonomy is central to moral theories and frameworks. The principle of autonomy (i.e., the principle that all individuals presumed to have decision-making capacity are afforded the right to self-determination, i.e., the freedom to make decisions for themselves) lies at the heart of various legal freedoms and rights, including freedom of speech and the right to privacy (or informational self-determination).

In applied ethics, the principle of autonomy has great practical relevance in medicine. The respect for a patient's autonomy is one of the most fundamental principles of medical ethics. In the field of computing, respect for the user's autonomy is an important issue as well, although it is frequently not labeled as such (as shown in Table 1). The title of the IFIP TC9 conferences, "HCC," refers to human choice, therefore to autonomy or self-determination, as a basic concern in the context of computing.

The relevance of the concept and the principle of autonomy in the field of computing can be explained by the trend toward increasingly "autonomous" machines, from the classical automation of repetitive tasks in manufacturing to the invisible control of complex sociotechnical processes in a (hypothetical) ubicomp world.

Starting from this perspective, I reviewed the discourse analysis [24] conducted on all ten HCC volumes [2, 4–6, 8, 11, 25, 26, 28, 29] and identified the main ethical issues connected to the topic of autonomy or self-determination. While the discourse analysis had mainly involved quantitative lexicometric methods, yielding histograms of words and of so-called n-grams (such as "working conditions" or "wireless sensor and actor networks"), my interpretation inevitably necessitated some qualitative contextual knowledge and is therefore not completely free of subjective judgment.

The result of my interpretation based on the discourse analysis of the HCC series is shown in the *left* column of Table 1. The *right* column lists related ethical issues specific to ubicomp that are mentioned in the three TA reports [9, 13, 18], each of them matched with its counterpart on the left side. The four issues under the "autonomy" umbrella—working conditions, virtual and augmented reality, privacy, and technology paternalism—are discussed in more detail in the following paragraphs.

Working conditions. Throughout the 1970s and 1980s, effects of computerization on employment, working conditions, and job satisfaction dominated the discourse at the HCC conferences [4, 25, 26, 29]. Participation of employees in management decisions became an issue, including the idea of participatory design processes for computer applications [4].

The issue of working conditions has returned in the ubicomp discourse, driven mainly by two aspects: the potential of ubicomp for close surveillance at the workplace and its tendency to blur the boundary between professional and private life [18]. The latter is also described in [20] as the "virtual merging of our social, family and working roles," forcing "new flexible boundaries between the different spheres of work, home and leisure, leading for some to a sense of increased stress and for others to greater empowerment" (p. 40).

Overall, the changes in working conditions because of computing have been discussed as a threat to human self-determination since the early days of computing; the original focus that led to the demand for participation in the design of the systems used at the workplace in the 1980s seems, however, to have lost importance in the ubicomp age. Instead, surveillance issues and around-the-clock availability of the workforce have become the new focal points of discussion.

Virtual and augmented reality. Communicating through virtual realities (e.g., provided by a computer game or a virtual working environment), taking on a virtual human role represented by an avatar, can be challenging because many natural aspects of communication may become unclear, for example, with whom we are communicating, who is following the communication, and how to secure virtual property [5, 6, 28].

In ubicomp, virtual or augmented reality techniques are likely to be used in a context connected to physical reality, such as remote medical diagnosis or surgery. There is a risk that communicative acts in such environments are more ambiguous than in a natural environment, which can cause damage, or that decisions are delegated to the technology in a way that affects the autonomy of the humans involved (both doctor and patient). On the other hand, augmented reality is expected to improve the precision of interventions and the availability of information during operations [13]. Similar arguments may apply in other safety-critical domains.

Ubicomp has shifted the focus of ethical concerns in the context of virtuality from the "within virtual worlds" perspective to the "real-world impact" perspective. This is not surprising, as ubicomp technologies are built to interact seamlessly with real-world processes via sensors and actuators. While in the early days of computing the discourse focused on how to keep control over virtual worlds (e.g., control over avatars or over virtual property), the ubicomp vision created more emphasis on issues of real-world processes controlled by humans and machines via virtual or augmented realities. The main issue here is the risks arising from potential damage caused by ubicomp systems, in particular in medical diagnosis and surgery.

This is linked to the issue of moral and legal responsibility for damage created by the use of computer systems (see Sect. 3.2).

Privacy. Privacy is an individual condition of life characterized by exclusion from publicness. In the context of computing, privacy is usually interpreted as "informational privacy," which is a state characterized "by controlling whether and how personal data can be gathered, stored, processed or selectively disseminated" [28, p. 58]. As an ethical issue in computing, information privacy is usually discussed as being threatened by computing infrastructures that facilitate the dissemination and use of personal data. The resulting requirement to protect individual privacy against data misuse entered many laws and international agreements under different terms, some of them focusing on the defensive aspect, such as "data protection," others emphasizing individual autonomy, such as "informational self-determination." This term first occurs in the HCC proceedings in 1986 [29], 3 years after the German Federal Constitutional Court declared the right to informational self-determination in its census verdict in 1983. At the same conference, "data protection" advanced to become one of the most frequently mentioned specialist terms. Threats against informational self-determination were mainly perceived as originating from governments. Later on, in the 2001 conference [28], the picture had changed in two respects: data protection was now—in the Internet age—discussed in connection with data security and encryption, and the focus had increasingly turned to the private sector. For example, the use of cookies, the creation (and sale) of profiles about individuals' financial behavior, and the private sector's interest in geographic data were discussed in the context of data protection in 2001 [28].

In the following conferences, the privacy discourse continued while integrating new and more specific issues, in particular biometric methods [8], health care (e-health) [5], and social media [11].

In the ubicomp discourse, the privacy issue revolves around three aspects:

- Automatic identification: Identifying persons even without their knowledge is much easier in a ubicomp world, because sensor data can easily be collected and combined [13, 18]. The discussion about automatic identification started with RFID [27], which is, however, less powerful than newer technologies of face recognition or device fingerprinting [18]. In a world of ubiquitous automatic identification, the amount of personal data generated and circulated is expected to increase dramatically [18].
- Location privacy: In addition to detecting an *agent*, ubicomp will usually generate data containing a reference to the *location of the action*. The aspect of location or positioning is linked to the general discussion about privacy in social networks [11] to the extent that social networking platforms will start tracking their users' locations automatically and in real time [18]. Location privacy is an important special case of privacy because public or private sector organizations that process location data can combine them into profiles from which not only the activities, but also the contacts of persons can be inferred [18, 19].
- Implications of transparency: In a ubicomp world, monitoring and recording virtually all processes and calculating indicators which are believed to represent

criteria relevant for making management decisions are feasible and affordable. The resulting "transparency" is not only a threat to privacy, but also to other aspects of self-determination: decisions may first be delegated to bureaucracy (indicator systems) and then from bureaucracy to computers (automated indicator systems), which means relinquishing autonomous decision making, or in fact ceding control to those who define the indicators [9].

The last concern mentioned above goes beyond privacy and will be revisited under the umbrella of responsibility (Sect. 3.2).

Technology paternalism. When someone believes they know the solution to someone else's problem and imposes this solution on that person even without their consent, this attitude is called "paternalism." There is a serious ethical dilemma behind paternalism: Imposing the solution violates the autonomy of the other person, whereas by not imposing it, one may not do the best possible thing in the other person's interest. As pointed out in [32], not only individuals, but any system, including governmental institutions and technical systems, can act in a paternalistic way. Paternalism can be "delegated" to machines by means of technology, and when executed by machines is called "technology paternalism" [32].

In the general discourse about the ethics of computing, paternalism is discussed mainly in two domains: security and e-health. It was implicitly addressed in the HCC 2002 proceedings [8], when anti-terror prevention measures introduced after the 9/11 attacks were discussed by asking the question whether "diminished liberty would be compensated by improved security" [8, p. 196]. In a similar way, technologies of biometrics such as fingerprinting, facial recognition, and iris scanning were discussed 10 years later at the HCC 2012 conference [11]. Paternalism was mentioned explicitly only in the context of e-health: While e-health can increase the autonomy of the patient who is empowered by information ("do-it-yourself healthcare"), doctors may make the "paternalistic decision" not to store important information if they know the patient will access it [5].[2]

Technology paternalism, however, is considered an inherent tendency in ubicomp systems, in particular when machine-learning techniques are applied to infer the user's intentions [9]. This thought is more clearly formulated in the IBM/ SwissRe/TA-SWISS study: "(The ubiquitous) computing environment will be unable to perfectly adapt to explicit requests or to correctly read the context or user intentions. New habits will therefore be acquired or 'tricks' to let the appropriate interface know what is desired, or even to cheat it in order to avoid undesired reactions. The systems will build user models, and the users will build their own approach to deal with them. The unpredictability and intended unobtrusiveness of the systems will make this a harder task for the user than before" [20, p. 40].

In health care, there is a special aspect of ubicomp raising serious ethical concerns: active implants and other remote methods of personal health monitoring [13]. The dilemma can be described as follows: On the one hand, the quality of life of patients

---

[2] Strictly speaking, this case is not about technology paternalism, but about an unintended consequence of introducing computational technology in a paternalistic environment.

who are chronically sick, undergoing rehabilitation, or at high risk can be improved by these technologies, in particular by reducing their dependence on hospital facilities. On the other hand, these opportunities will be accompanied by the risk that active implants might have unexpected side effects or, viewed from a more general perspective, that an "over-instrumented" way of practicing medicine might have a negative psychological impact on patients subjected to close observation [13].

Another important aspect of technological paternalism discussed in the ubicomp context is the use of tracking and tracing devices in dependency relationships. On the one hand, tracking can enhance the safety and security of the tracked persons, in particular patients, children, or employees. On the other hand, tracking represents a serious threat to the self-determination the tracked individual. Who should be given the right to track and trace whom for what purpose? [18].

3.2 Responsibility

Computing professionals work in environments where small causes can have large effects. Decisions made and actions taken during software development may have serious consequences in practical application, as in the famous case of the Therac-25 radiation therapy machine that killed several patients by giving massive overdoses of radiation.

The "small cause—large effect" property of digital technology leads to questions of who is responsible, both in legal and in broader moral terms, for damage that may result from using computer systems. This "attributional" concept of responsibility is also known as "accountability" because it addresses the question of who is accountable for the effects of a chain of actions. In the case of the production and use of computer hardware and software, attributing legal and moral responsibility is difficult due to a problem that has been termed "the problem of many hands" [1].

A different concept of responsibility is social responsibility, which addresses the obligation of an individual or an organization to act with the goal of benefiting society at large.

Legal and moral responsibility. As early as 1980, the issue of who will be responsible for "computer decisions" and "decisions based on wrong information" in an increasingly automated world was discussed on the second HCC conference [25]. A change in the public perception of computers was reported: "The public conviction of objectivity of computer decisions has given way to a feeling of the irresponsibility of such decisions" [25]. This issue recurred later in a critical discussion of the agent concept: "The delegation of any task to a software agent raises questions in relation not only to trust but also to its autonomy of action and decision, and to the location of responsibility, both moral and legal, for the outcomes of those decisions and actions" [8].

The issue of responsibility for decisions delegated to machines was discussed in the context of professional responsibility, which was defined as "a kind of responsibility that combines traits of legal and of moral responsibility" of the IT professional for the outcome of decisions taken [5].

The application context in which the issue of legal and moral responsibility was discussed shifted from e-health in 2006 [5] to social media in 2012 [11]. In the social media context, the responsibility of the *user* was addressed for the first time: "The people who communicate via social media are morally responsible for that communication and for the foreseeable effects of it. This responsibility is shared with other people who have affected and contributed to that communication as part of a sociotechnical system. This identifies moral responsibility both for those who create the message for its unintended but foreseeable effects, and for those who use a system to wrongfully harm others" [11].

In the ubicomp discourse, the issue of responsibility for decisions made by (increasingly autonomous) computer systems is a central concern. A "basic ambivalence" of ubicomp applications is seen in their impact on human control: Will we gain more control over our environment in a ubicomp world, or will the autonomous systems start to control us? [9]. When the systems make decisions that turn out to be against the user's intention, it will be difficult to attribute responsibility: "The penetration of everyday life with systems whose behavior is dependent on complex hardware and software in a distributed system makes it quite difficult to identify the cause and causer where harm occurs. This situation could be further exacerbated (...) because there will be a very great incentive to use (...) programs acting on behalf of their users (software agents). The incentive arises from the fact that the flood of possibilities, in conjunction with the social pressure also to use them, is pushing the boundaries of human processing capacity" [13]. The basic problem with regard to responsibility is the fact that machines are not capable of making commitments, leading to a problem called "dissipation of responsibility"[3]: "A promise made by a machine—e.g. to carry out a particular function—is in principle worthless as it cannot feel obligation and cannot be held responsible. The inability of machines to make commitments in principle excludes them from social interaction. Consequently, there is a danger of a 'dissipation of responsibility' (...A) fine distribution of cause and responsibility as a result of the multilayered or networked nature of digital ICT can arise which can no longer be controlled by legal means" [13, p. 265]. However, other authors emphasize that this technology can improve accountability in organizations [7].

To conclude, the ubicomp vision has highly magnified one aspect of the accountability issue already established in the ethics of computing discourse: the implications of increasingly autonomous machines for moral and legal responsibility. These implications are complex, and there is no single standard that could be applied to all potential applications.

---

[3] Dissipation of responsibility addresses the problem of accountability in complex distributed human–machine systems [16, 30] and is not to be confused with what is known as "diffusion of responsibility." This is the psychological phenomenon that people tend to feel less responsible for their individual actions the larger the group of people who could take action.

Social responsibility. Social responsibility differs from moral and legal responsibility (or accountability) discussed above by addressing an obligation to act toward the benefit of society, regardless whether one is accountable for the outcome of an action. In the HCC discourse, social responsibility was first discussed as an obligation on the part of large companies and the public to pay attention to the negative social impacts of an ongoing new wave of (computer-based) industrial automation [26].

After automation in the 1970s, globalization was recognized in the 1980s as an emerging aspect of computerization that should be dealt with in a socially responsible way: "Because of the marriage between computer technology and telecommunications the globe has shrunk to the size of a ping-pong ball, crowded with our traditional unsolved problems" [29]. In particular, "multinational corporate social responsibilities" of the mainly US-based computer industry were discussed. [29]. More than 10 years later, the contributions of information systems to the transparency of business organizations [5] and to corporate social responsibility (CSR) entered the discourse [11].

Government policies related to new opportunities and risks of computing were discussed as well in the context of social responsibility, such as national policies related to the role of computers in nuclear weapons systems (including President Reagan's proposed Strategic Defense Initiative, known as the "Star Wars Program") [29], the introduction of national identification schemes after the 9/11 attacks [8], and policies related to new critical infrastructures [11]. In 1990, technology assessment was discussed as an approach for governments to implement social responsibility in the use of new technologies [4].

Besides companies and governments, the individual IT professional has been addressed by the issue of social responsibility throughout the HCC discourse. In 1980, having a sense of social responsibility still seemed to counter a widespread prejudice: "It is sometimes said that computer—and other—specialists do not appreciate the social effects of their activities" [25]. In the following years, IFIP TC9 became instrumental in motivating, facilitating, and reflecting the development of ethics codes of national computer professional associations around the world [2–6, 8, 11, 28, 29], a process that cannot be reported in detail in this article.

In the ubicomp discourse, there is one additional aspect of social responsibility already mentioned in Sect. 3.1, namely the potential consequences of transparency on automated decision making: Is it socially responsible to allow the diffusion of technologies that could replace human choice with the automated application of indicators and routines defined by a few people [9]?

3.3 Distributive Justice

Distributive justice concerns the allocation of goods (wealth, opportunity, respect) in society and is linked to issues of equality, power, need, responsibility, and other basic concepts discussed in ethics. Ethics in computing relates to two specific issues of distributive justice: the digital divide and sustainable development.

Digital divide. In the HCC conferences, the term "digital divide" first occurred in the 2002 proceedings [8]. The issue as such, however, was discussed at earlier conferences using different terms, such as "the information-rich" versus "the information-poor" [29] and "computer literacy" [4, 8, 29], in the context of information technology and developing countries [29] as well as technology transfer [4], in terms of "digital inclusion" versus "digital exclusion" [2], and finally, as one aspect of intellectual property and the phenomenon of piracy [6].

In the three ubicomp studies, the digital divide was mentioned only in [13], defined here as "the jeopardization of social justice through the division of society into those who have access to the information society and those who are excluded" (p. 41). This study assigns a high probability to the scenario that the digital divide will be reduced by the availability of better user interfaces and the continued diffusion of ICT, a hypothesis that has at least partly become reality through the spread of the mobile phone around the globe as well as programs providing affordable computers to schools in developing countries [33].

Sustainable Development. The aim of sustainable development can be defined as solving a double problem of distributive justice, namely both intergenerational and intragenerational justice [15, 21].

First mentioned at the 1998 HCC conference [28], the relationship between the aim of sustainable development and the information society (or knowledge society) was discussed in 2002 [8] and more broadly in all three succeeding conferences [2, 5, 6]. The 2012 proceedings [11] contain a surprisingly high number of "sustainable X" terms, such as "sustainable innovation," "sustainable business," "sustainable growth," "sustainable computing," "sustainable consciousness," and "sustainable governance" [11], whose relation to the concept of sustainable development is not always clear. The term "sustainable development" itself had almost vanished in the 2012 proceedings.

In the ubicomp discourse, the issue of sustainable development was addressed in several ways. First, ubicomp technologies were attributed a higher dematerialization potential (potential to replace physical goods and processes by virtual ones)[4] compared to traditional computing, thus creating opportunities for sustainable development [13, 16]. Second, the chemical elements (covering half of the periodic table) needed to produce the small ubicomp devices in vast numbers and the increasing problem that they are not recycled[5] were mentioned as a threat to sustainable development [12, 13, 31]. In addition, the risks of an emerging new critical and vulnerable infrastructure, raising questions of the distribution of safety in society, were mentioned in [18] with regard to positioning technologies: "They are becoming new critical infrastructures the malfunctioning or collapse of which can have far-reaching consequences" (p. XXI).

---

[4] Dematerialization effects of computing result in relevant reductions in energy consumption and greenhouse gas emissions, if not compensated by rebound effects [10, 17].

[5] Embedded ICT hardware can also compromise established recycling processes [22, 23, 34].

Ubicomp seems to be ambivalent with regard to sustainable development; this is also true of computing in general [14], but the connection to physical and ecological aspects can be seen more clearly in the case of ubicomp.

4 Conclusion

Viewed against the background of the general discourse on ethics in computing as it has evolved over four decades in the HCC conferences of IFIP TC9, most of the ethical issues discussed in the ubicomp discourse—as far as it is reflected in the three studies—turn out to be special cases of persistent ethical issues of computing, but with some new aspects that were not anticipated in the earlier discourse. These new aspects are as follows:

- the potential for closer surveillance and around-the-clock availability of employees;
- virtual realities having direct effects on physical realities in safety-critical domains, such as e-health;
- ubiquitous automatic identification and its implications for informational self-determination, including location privacy;
- complete transparency of processes creating incentives to automate indicator-based decisions;
- technology paternalism in health care and other domains where dependency relationships exist, such as parenting;
- legal and moral responsibility (accountability) of autonomous computer systems and the "dissipation" of responsibility;
- opportunities to overcome digital divides or facilitate digital inclusion;
- sustainable use of natural resources, conservation versus dissipation of materials;
- emergence of a new critical infrastructure and the social distribution of safety.

Designers of ubicomp technology should take these aspects into account and consider their complex ethical implications when developing applications. Decision makers in organizations introducing such applications should be aware of their responsibility for the ethical implications of the technology.

References

1. Ad Hoc Committee for Responsible Computing: Moral Responsibility for Computing Artifacts: Five Rules, Version 27. https://edocs.uis.edu/kmill2/www/TheRules/ (2010)
2. Avgerou, C., Smith, M.L., Van den Besselaar, P. (eds.): Social Dimensions of Information and Communication Technology Policy. In: Proceedings of the Eighth International Conference on Human Choice and Computers (HCC8), IFIP TC 9, Pretoria, South Africa, 25–26 September, Springer (2008)

3. Berleur, J., Brunnstein, K. (eds.): Ethics of computing: codes, spaces for discussion and law: a handbook prepared by the IFIP ethics task group. Chapman and Hall, London (1996)
4. Berleur, J., Drumm, J. (eds.): Information technology assessment. In: Proceedings of the 4th IFIP TC9 International Conference on Human Choice and Computers, HCC-4, held jointly with the CEC FAST Programme, Dublin, Ireland, 9–12 July 1990, North-Holland (1991)
5. Berleur, J., Nurminen, M.I., Impagliazzo, J. (eds.): Social informatics: an information society for all? in remembrance of Rob Kling. In: Proceedings of the Seventh International Conference on Human Choice and Computers (HCC7), IFIP TC 9, Maribor, Slovenia, 21–23 September, Springer (2006)
6. Berleur, J., Hercheui, M.D., Hilty, L.M. (eds.): What Kind of Information Society? Governance, Virtuality, Surveillance, Sustainability, Resilience—9th IFIP TC 9 International Conference, HCC9 2010 and 1st IFIP TC 11 International Conference, CIP 2010, Held as Part of WCC 2010, Brisbane, Australia, September 20–23, Springer (2010)
7. Boos, D., Guenter, H., Grote, G., Kinder, K.: Controllable accountabilities: the internet of things and its challenges for organisations. Behav. Inf. Technol. **32**(5), 449–467 (2013). doi:10.1080/0144929X.2012.674157
8. Brunnstein, K., Berleur, J. (eds.): Human choice and computers. In: Issues of Choice and Quality of Life in the Information Society, IFIP 17th World Computer Congress, TC9 Stream, 6th International Conference on Human Choice and Computers: Issues of Choice and Quality of Life in the Information Society (HCC-6), 25–30 August, Montreal, Quebec, Canada. Kluwer Academic (2002)
9. Kündig A., Bütschi, D. (eds.): Die Verselbständigung des Computers. vdf Verlag, ETH Zürich (2008)
10. Erdmann, L., Hilty, L.M.: Scenario analysis: exploring the macroeconomic impacts of information and communication technologies on greenhouse gas emissions. J. Ind. Ecol. **14**(5), 826–843 (2010)
11. Hercheui, M.D., Whitehouse, D., McIver W.J., Phahlamohlaka, J. (eds.): ICT Critical Infrastructures and Society. In: 10th IFIP TC 9 International Conference on Human Choice and Computers, HCC10 2012, Amsterdam, The Netherlands, 27–28 September, Springer (2012)
12. Hilty, L.M.: Electronic waste—an emerging risk? Environ. Impact Assess. Rev. **25**(5), 431–435 (2005)
13. Hilty, L.M., Behrendt, S., Binswanger, M., Bruinink, A., Erdmann, L., Froehlich, J., Koehler, A., Kuster, N., Som, C., Wuertenberger, F.: The Precautionary Principle in the Information Society—Effects of Pervasive Computing on Health and Environment. Second Revised Edition. Edited by the Swiss Center for Technology Assessment (TA-SWISS), Bern, Switzerland (TA46e/2005) and the Scientific Technology Options Assessment at the European Parliament (STOA 125 EN) (2005)
14. Hilty, L.M., Lohmann, W.: An annotated bibliography of conceptual frameworks in ICT for sustainability. In: Hilty, L.M., Aebischer, B., Andersson, G., Lohmann, W. (eds.) ICT4S 3013: Proceedings of the First International Conference on Information and Communication Technologies for Sustainability, ETH Zurich, 14–16 February, 288-300. E-Collection ETH Institutional Repository (2013). doi:10.3929/ethz-a-007337628
15. Hilty, L.M., Ruddy, T.F.: Sustainable development and ICT interpreted in a natural science context: the resulting research questions for the social sciences. Inf. Commun. Soc. **13**(1), 7–22 (2010)
16. Hilty, L.M., Som, C., Köhler, A.: Assessing the human, social and environmental risks of pervasive computing. Hum. Ecol. Risk. Assess. **10**(5), 853–874 (2004)
17. Hilty, L.M., Köhler, A., von Schéele, F., Zah, R., Ruddy, T.: Rebound effects of progress in information technology. Poiesis and Praxis: Int. J. Technol. Assess. Ethics Sci. **1**(4), 19–38 (2006)
18. Hilty, L.M., Oertel, B., Wölk, M., Pärli, K.: Lokalisiert und Identifiziert. Wie Ortungstechnologien unser Leben verändern. vdf Verlag, ETH Zürich (2012)

19. Hilty, L.M., Oertel, B., Wölk, M., Pärli, K.: Locating, tracking and tracing: from geographic space to cyberspace and back. In: Michalek, T.C., Hebakova, L., Hennen, L., Scherz, C., Nierling, L., Hahn, J. (eds.) Technology Assessment and Policy Areas of Great Transitions, pp. 336–341. Informatorium, Prague (2014)

20. IBM Zurich Research Laboratory, Swiss Re Centre for Global Dialogue, TA-SWISS: Exploring the Business and Social Impacts of Pervasive Computing. Zurich (2006)

21. Isenmann, R.: Sustainable information society. In: Quigley, M. (ed.) Encyclopedia of Information Ethics and Security, pp. 622–630. IGI Global, Hershey (2008)

22. Köhler, A.R., Hilty, L.M., Bakker, C.: Prospective impacts of electronic textiles on recycling and disposal. J. Ind. Ecol. **15**(4), 496–511 (2011). doi:10.1111/j.1530-9290.2011.00358.x

23. Kräuchi, Ph, Wäger, P., Eugster, M., Grossmann, G., Hilty, L.M.: End-of-life impacts of pervasive computing. IEEE Technol. Soc. Mag. **24**(1), 45–53 (2005)

24. Lignovskaya, E.: Human Choice and Computers 1974–2012. Eine Diskursanalyse mit Hilfe lexikometrischer Verfahren. Facharbeit im Fach Wirtschaftsinformatik am Institut für Informatik. Universität Zürich (2013)

25. Mowshowitz, A. (ed.): Human choice and computers 2. In: Proceedings of the Second IFIP Conference on Human Choice and Computers, Baden, Austria, 4–8 June 1979, North-Holland (1980)

26. Mumford, E., Sackman, H. (eds.): Human choice and computers. In: Proceedings of the IFIP Conference on Human Choice and Computers, Vienna, 1–5 April 1974, North-Holland (1975)

27. Oertel, B., Wölk, M., Hilty, L.M., Köhler, A.: Security aspects and prospective applications of RFID systems. Bundesamt für Sicherheit in der Informationstechnik, Bonn (2005)

28. Rasmussen, B.L., Beardon, C., Munari, S. (eds.): Computers and networks in the age of globalization. In: IFIP TC9 Fifth World Conference on Human Choice and Computers, pp. 25–28 August 1998, Geneva, Switzerland. Kluwer Academic (2001)

29. Sackman, H. (ed.): Comparative worldwide national computer policies. In: Proceedings of the Third IFIP TC9 Conference on Human Choice and Computers, Stockholm, Sweden, 2–5 September 1985, North-Holland (1986)

30. Som, C., Hilty, L.M., Ruddy, T.: The precautionary principle in the information society. Hum. Ecol. Risk. Assess. **10**(5), 787–799 (2004)

31. Som, C., Hilty, L.M., Köhler, A.R.: The precautionary principle as a framework for a sustainable information society. J. Bus. Ethics **85**(3), 493–505 (2009)

32. Spiekermann, S., Pallas, F.: Technology paternalism—wider implications of ubiquitous computing. Poiesis Praxis: Int. J. Technol. Assess. Ethics Sci. **1**(4), 6–18 (2006)

33. Streicher-Porte, M., Marthaler, C., Böni, H., Schluep, M., Camacho, A., Hilty, L.M.: One laptop per child, local refurbishment or overseas donations? sustainability assessment of computer supply scenarios for schools in Colombia. J. Environ. Manage. **90**, 3498–3511 (2009). doi:10.1016/j.jenvman.2009.06.002

34. Wäger, P., Eugster, M., Hilty, L.M., Som, C.: Smart labels in municipal solid waste—a case for the precautionary principle? Environ. Impact Assess. Rev. **25**(5), 567–586 (2005)

Socio-ethical Issues of Ubicomp: Societal Trends, Transparency, and Information Control

Katharina Kinder-Kurlanda and Daniel Boos

Abstract In this paper, we undertake a consideration of the changes that are occurring in workplaces with the rise of ubiquitous computing (ubicomp). We offer two case studies from research projects to show how contemporary societal trends —such as an increasing attention toward audit, transparency, and control over complexity—play out in concrete workplace settings. Employees and employers discussed ubicomp technologies within the context of the data they would provide and the transparency and control over complexity which this data promised. Companies and even whole industries were seen to be under pressure to become transparent and to provide proof and accounts of everyday work activities. Ubicomp technologies could then lead to new accountability challenges as control shifts could occur with the new availability of data in places about which previously little information had been available. From our case studies, we suggest that the ethical challenges that those face who are tasked with deciding whether to introduce ubicomp technology and also the ethical dilemmas that can occur for those who use such technologies are connected to wider societal trends of informational ubiquity. We find ourselves in socio-technically produced audit cultures in which control over information is all-important. Decision takers within organizations and those regulating the field of intelligent work environments should take into account the results of an increased transparency that the new technologies provide; how liability issues come into play; how responsibilities may shift with the new system; and how to ensure that actors' control capabilities over the situations they find themselves in are sufficient.

Keywords Ubiquitous computing · Intelligent workplace environments · Internet of things · Ethics · Socio-ethical issues · Ethnography

K. Kinder-Kurlanda (✉)
GESIS—Leibniz Institute for the Social Sciences, Unter Sachsenhausen 6-8, 50674 Cologne, Germany
e-mail: katharina.kinder-kurlanda@gesis.org

D. Boos
Swisscom (Schweiz) AG, 3050, Bern, Switzerland
e-mail: boos@pong.ch

© Springer International Publishing Switzerland 2015
K. Kinder-Kurlanda and C. Ehrwein Nihan (eds.), *Ubiquitous Computing in the Workplace*, Advances in Intelligent Systems and Computing 333, DOI 10.1007/978-3-319-13452-9_5

1 Introduction

Contemporary workplaces may often look like what we used to think of as work-places—people still sit at desks in offices, work on construction sites or in shops, and, considering the global picture, many employees perform tasks similar to those of people in comparable trades before the age of the internet. At the same time, we witness work and its organization having become entangled with information technology in such a way that they often cannot even be told apart any longer [1]. Workplaces are the target of developments in information technology that aim to increase productivity and to open new service opportunities. This fact also brings with it many ethical challenges, especially as computers are becoming mobile and ubiquitous and have begun to disappear to the human eye. Concurrently, computers in workplaces are becoming part of increasingly networked structures and global information flows in which data may end up with unwanted or unintended recipients. Unforeseen consequences may develop for individuals who may not have been aware that they were part of generating certain data. The situation of ubiquitous computing (ubicomp) in the workplace is therefore hard to survey with regard to consequences to stakeholders, which makes it difficult for such stakeholders to take decisions about whether to introduce, how to implement, and how to use such technologies.

In addition to the fact that contingencies of ubiquitous systems targeting workplaces are often as of yet unforeseeable, legal frameworks to guide decision making about technology implementation and design also were not set up with the new technological developments in mind and in fact new regulatory approaches become necessary [2]. Hence, decisions often hinge on ethical considerations. As ubiquitous computing is becoming part of people's workplaces, many ethical issues seem obvious, as problems concerning the power balance within the organization, humane working conditions, surveillance, and last but in no means least, the right to privacy come into play [3–5]. Such issues put at stake socio-ethical regulations and axiological frameworks that go beyond individual workplaces or even individual companies. Handling these challenges should therefore not solely be assumed on the level of individual actors' personal ethical obligations [6]. Policy making, however, also faces challenges as new ways may need to be found to influence government or company policies with public debates increasingly taking place in networked spaces [7]. If a wider consideration of intelligent workplaces is to happen, however, we need to understand the changes that are occurring with ubiquitous computing technologies in workplaces on a more general level in order to be able to identify some overall ethical concerns.

A prerequisite to generalizing ethical concerns of ubicomp is to be clear about and gain a comprehensive understanding of how the technologies influence their environments. Many concepts in the past years have questioned the idea of allo-cating causes for change either merely within technologies or merely within the environment but rather describe complex socio-technical arrangements as scholars increasingly call into question analytical efforts to treat technology, work, and organization as distinct conceptual units. Approaches range from proposing a symmetry between non-humans and humans to study translation processes and heterogeneous networks [8], to a concept of affordances looking at how materiality of an object "favors, shapes, or invites, but at the same time constrains a specific set of uses," [9] to emphasizing the constitutive entanglement of technology, work, and organization and the intertwining of human and technology in the form of socio-material practices [1, 10, 11].

In essence, following these insights, we adopt the following model of innovation: Every technology emerges under the specific social and cultural conditions of its time and place, but in its emergence already changes these conditions. Technology and its context (work and organization) are entangled to a degree where they are not only indistinguishable but in fact inseparable ontologically, so that it has become difficult to think of them as separate entities. We therefore consider *technology assemblages* [12] and *actor-networks* [8], which comprise not only technological artifacts but also other human and non-human actors and structures.

We also need to consider the relationship between technology and society in more general terms if we want to identify areas in which new arising ethical challenges require legal or regulatory consideration in order to not de facto leave decisions merely to individuals' personal ethical obligations. Societies evolve with new technologies becoming available, and technologies in turn evolve as societies change [13, 14]. The socio-technical arrangements that we witness in organizations are part of a bigger picture and reflect the historical conditions of a certain time and place. The bigger picture of societal "trends" [15] and changes thus provides a context for the ethical challenges that people face in intelligent workplaces. There are some indications provided in the literature of what these bigger changes might entail—diagnoses include proclaiming the rise of control via information [16] and analyzing the power of algorithms in an age of big data: Metadata analyses are seen to allow not only to understand individual positions but to gauge or even predict tendencies in mass behavior [17, 18]. Other authors describe an increasing auto-mation of knowledge work [19] and the rising need for employees with strong computational skills resulting in a polarized job market [20]. Observing how the bigger picture of societal trends plays out in actual workplaces is necessary in order to learn more about the ethical challenges of ubicomp in workplaces. Such observations can in turn also contribute to a better understanding of the bigger picture. Following ideas in sociology and anthropology, individuals are not only influenced by the societies they live in [21] but there is also a cultural flow "back" in the sense that individuals constantly produce meaningful external forms that then are observed by others and hence contribute to changing culture and society [22].

In this paper, we offer two case studies as a first step toward an observation of societal trends as they play out in workplaces. We deem it necessary to do so in order to understand what needs to be taken into account when thinking about the ethics of ubicomp in workplaces. Both case studies are summarizing accounts of detailed long-term field studies that the authors were involved in, more detailed results of which have been published elsewhere [23–25]. Both projects employed qualitative methods such as interviews and observation to study workplaces in industry in which managers and workers were faced with new, intelligent workplace environments. With our goal of contributing first steps toward gauging the societal impact of ubicomp in order to identify areas in which new ethical challenges require consideration, we focus in this paper on the reasons why actors pushed for the introduction of these technologies, on their argumentations for or against introduction and how we found these to be influenced by existing theories and ideas about technological ubiquity prevalent at the time of enquiry. Argumentations were therefore not only linked to individual experiences in workplaces and organizations but were also connected to the more general promises of intelligent work environments and ubicomp as they were, for example, also seen in computing and business studies such as that ubicomp will improve business processes, for example, by speeding up the handling of goods, enhancing products by making them smart, or developing new services [26, 27].

2 Intelligent Workplace Environments Enabled by Ubicomp

The first of our two cases is taken from a long-term ethnographic study of construction and maintenance workplaces in the UK between 2005 and 2009 within the context of the ubiquitous computing research project "Networked Embedded Models and Memories of Physical Work Activity" (NEMO) at Lancaster University. We assumed that by employing ethnography we might be able to analyze and describe work processes within the organization in such a way as to inform design. However, during the ethnographic fieldwork, it became apparent that this initial premise did not capture the overwhelming impact both high-level managerial strategic thinking and general attitudes of workers and managers had on decisions made about the technologies. We therefore in the following focused on the more broadly formulated question of why ubicomp technologies had become necessary in order to remain competitive and successful. More detailed questions asked were as follows: What was the context for the decision to introduce the technologies and how had they become necessary in order to remain competitive and successful? What were the changes in society that made these technologies important and how in turn were the technologies part of establishing these changes? The research took into account individual workplaces, changes within the organization, and the major discourses at the time as they were reflected in both workplaces and organization.

The second case is taken from a study of workplaces employing semi-structured interviews and participant observation conducted within the context of the "Stop Tampering of Products" (SToP) project. The SToP project ran from 2007 to 2009 and had the aim to develop knowledge of how to prove the genuineness of pharmaceutical, luxury goods and other industrial products throughout the supply chain by means of a ubicomp application. The aim of our qualitative inquiry was to understand how ubicomp technology, organizations, and humans mutually constitute each other in the ongoing sociomaterial practices we witnessed. In particular, we were interested in how the capacities of a technology can constrain or enable actions of the people in the organization. The focus on constraints and enabling was motivated by our aim of contributing some answers to the research question of what changes within the organizations with the introduction of ubicomp technologies. What organizational issues was ubicomp connected to, especially with regard to its contribution to the organization, alignment, and transition of individual work tasks and skills? The research took into account individual workplaces and related them to the surrounding, changing organization.

2.1 Case 1: UK Road Construction

From our study of UK road construction and maintenance workplaces, we here focus on two topics which we found to feature most prominently in the various actors' accounts of the reasons, thinking processes, and discussions about why ubicomp technologies had become desirable in order to remain competitive and successful as an organization. These topics are the related issues of improving "transparency" and of managing liability issues. We aimed to trace how these two topics were relevant within the context of broader changes that the technologies were entangled with (according to both our interviewees' accounts and informed by analyses from the scientific literature) and the ethical considerations that therefore should be taken into account.

2.1.1 Transparency

Within the road construction and maintenance sector in the UK, we found a particularly complex arrangement of contracts and subcontracts. The Highways Agency encouraged several larger companies to compete in bidding for highway contracts. Local councils would also employ private contractors for maintenance and construction work (which they had previously performed themselves), but not all councils had outsourced all maintenance and construction work and when such outsourcing did arise, it could involve multiple companies. In addition, companies then would employ subcontractors for specific types of work. This situation became even more complicated as contracts, and therefore, responsibilities and alliances changed over time and also because of other schemes. Government initiatives such as the public private partnerships[1] and resulting Design-Build-Finance-Operate (DBFO) schemes since 1996[2] had created exceptions and special circumstances for specific types of work. For example, a company might be maintaining one specific

---

[1] http://www.hm-treasury.gov.uk/ppp_index.htm (June 13, 2014).

[2] http://www.highways.gov.uk/roads/2992.aspx (June 13, 2014).

stretch of a motorway under a DBFO scheme independently of the usual contracts in the area. The complex situation had led to shared responsibilities for work being carried out, which became hard to oversee and manage.

Consequently, companies were seen to be under pressure from clients, insurance companies, and regulatory bodies to become more "transparent". Increasingly, companies were required to provide proof of work having been carried out and to give details about when, by whom, and under what circumstances this had occurred. Due to their capacity to record assets' location and to collect data about work activities, ubicomp technologies promised to provide the tools for providing data and therefore "transparency" in exactly such complex situations, which was seen as necessary in order to manage and control them. New sensors, global positioning systems (GPS), and radio frequency identification (RFID) were able to collect digital data in places previously only accessible to non-digital (e.g., based on handwritten records) data capture and thus met the demand for a specific kind of transparency and control over information.

While the situation within the road construction sector may be specific, contemporary organizations increasingly operate under conditions of complexity with organizational boundaries being hard to define. The call for transparency and the resulting resorting to ubicomp technologies in the face of multi-stakeholder complexities is certainly transferable to other domains such as supply chains [28], animal husbandry [29], or retail [30].

Within our setting, an example of managing pressures imposed by transparency demands with the help of ubicomp technology was the introduction of GPS tracking systems into vehicles. At the time of our study, using GPS for tracking vehicles was a relatively new technology which not all companies employed. For example, at a company we looked at, a GPS tracking system for gritting vehicles had only recently been introduced. GPS units in the vans were transmitting location data to a central server. The aims of this system were all connected to gathering data in the previously inaccessible domains of vehicles and traffic. The company expected to be able to track stolen vehicles, to prove to customers where, when, and what type of work had been carried out, and to have some data about vehicle speed and location in the case of accidents. These capacities were driven by demands for transparency from various actors: For example, clients expected the improved service of being provided with information on the status of contracted work and vehicle insurance companies rewarded GPS tracking of vehicles. In this way, demands for transparency posed to the organization from the outside led to a considerable change in the workplaces, e.g., of the gritting vehicle drivers, making their activities visible and "transparent." Potential implications for the drivers were enhanced surveillance and control of their work through the middle management and the possibility to prove innocence or guilt in the case of incidents such as a suspicion of breaking the speed limit or being involved in an accident. However, transparency demands could also stem from inside the organization.

An example of managing internal transparency demands with the help of ubicomp technology in the setting of road construction was the growing convergence of new data, i.e., data gathered by systems such as vehicle cameras on cars and highways and GPS systems. All these data converged in traffic control rooms that facilitated continuous and timely monitoring and improved management of traffic, work, and incidents on all high speed roads in a specific area. Control rooms organized the response management and the road space booking system for construction or maintenance work. Wall mounted screens would show live feeds from stationary motorway CCTV cameras, and smaller screens showed the feed of other cameras such as road construction supply depots' security cameras. Cameras could be controlled remotely, and control center personnel were able to zoom in on areas of interest. Such area control rooms received calls from bigger, regional control centers, and the police, and if there was an incident, they would send out response vehicles. All response vehicles were monitored by a GPS location system that reported their location to the control room where units were visualized on a map. There was also a log so that vehicles' movements could be "replayed" on the map. In case of an incident, the control room could determine which vehicles were closest and then send those to a specific location. However, the first action would often be to call the response units responsible for the specific area on their mobile phone. In fact, most of the communication was done by mobile phone, and the GPS and cameras mostly served as a backup system. Control center employees were very enthusiastic about the benefit of data gathered by cameras and GPS systems as it allowed controlling all vehicles and work going on an area. They aimed to obtain as much data as possible and were very keen to extend monitoring to areas as of yet inaccessible to data capture, such as motorways not yet fitted with cameras or vehicles not yet fitted with GPS tracking systems. At the time of our study, areas and assets not yet available to electronic data capture were, for example, the general foremen vehicles, which sometimes would be able to get to an incident quicker than the response unit vans. One control center employee said: "You can actually see what is going on" and stated that this was "such a benefit."

This description is a snapshot of the system from about five years ago at the time of writing this text, and we can assume that by now, the technological capacities of monitoring roads and vehicles and both type and amount of data converging in control rooms have grown considerably. The point being made here, however, is that there are also demands for transparency from within the organization, and many ideas by various actors on how and why ubicomp technologies may increase transparency so as to be beneficial for the tasks that they need to perform in the line of their work.

2.1.2 Liability

A more indirect way in which demands for transparency were carried into the organization was the increased importance of liability, which also made data capture in previously inaccessible domains interesting and beneficial. For example, in the case of the NEMO project, which aimed to improve health and safety of workers by accurately measuring tool usage times and exposure to vibration directly on work sites, we found that the health and safety context was very strongly defined by liability risks [23, 31]. While there were obviously genuine concerns about health and safety in the risky environments of road work sites, health and safety also needed to be seen within the context of what many managers described as a rise in a "blame culture". Companies were increasingly becoming a target for insurance and other legal claims. In addition, the privatization of what were formerly public sector services increased the visibility of companies as targets for claims. In order to remain efficient, companies had to find ways to prevent lawsuits, or at least be able to give proof that they had fulfilled their legal obligations.

Within the fieldwork, three main areas were identified in which liability played an important role in the industry. First, health and safety in workplaces could be enforced to avoid legal cases brought against employers by injured employees. Second, civil lawsuits would increasingly (at least this was managers' and administrators' perception) be brought against companies by members of the public. Managers and operatives told us that they had experience with drivers wrongfully suing the company for compensations for damage to their cars. Third, companies had to deal with liability issues when they had to prove fulfilling legal obligations to fulfill client contracts.

Several managers told us that they held a growing "claims culture" or "blame culture" in the UK responsible for the increase in (wrongful) suits. One manager said: "...we've very, very much got a claims culture in this country, which is frightening." Managers would also see the increasing blame culture as a reason why health and safety was even more important as accidents became even costlier through follow-up lawsuits. For example, one operations manager said: "It's a culture thing, that they sue you at the drop of a hat (...). If you cut corners, not only for the safety of your operatives but also for the safety of the public you tend to find that it'll be counterproductive because if you get sued for tripping somebody up on a footpath and they break their arm, you can guarantee you lose 10,000 pound (...). So if you understand that safety goes hand in hand with making money you can't divorce them."

Managers also believed that as a private company they were more exposed. One manager said: "Contractors are far easier to be shot at by the HSE[3] in terms of prosecution and put in the spotlight than the county council, because the county council, you've got policy making committees and this, that and the other, so you can never actually pin it down to one individual, and say, this is the person who ultimately started it."

The demands of liability were also prevalent in the example of GPS tracking of vehicles. The system at the company in question was used to confirm in a court case that a driver, who had been involved in an accident, had not been driving over the speed limit.

Data collected with the NEMO technology in the project was hoped to be useful for providing information about adherence to health and safety rules which was

---

[3] HSE = Health and Safety Executive, a watchdog agency in the UK (http://www.hse.gov.uk).

thought to be useful in the case of legal claims by employees with health problems due to long-term exposure to vibration or noise.

To summarize, in the NEMO project, we saw how ubicomp technologies collected digital data in places previously only accessible to non-digital data capture and thus met the demand for specific kinds of transparency and control over information, which had become especially necessary in order to mitigate liability risks. The ways in which opportunities and restrictions through suddenly available data would play out were diverse, with both benefits and drawbacks for individual actors being possible and sometimes unforeseeable. It became clear, however, that ubicomp technologies were discussed within the context of the data they would provide and the transparency and control over complexity which this data promised. In the concrete workplace settings, we could observe the wider societal trend toward more transparency and the perceived rise of a "claims culture" play out and provide a background for why and how technologies would be used. It was against this background that ethical dilemmas would occur and need to be solved, which is something we will explore in the second case study.

2.2 Case 2: Tracing Luxury Goods

From our study of the attempt to trace luxury goods, we here focus on two topics, which we found to feature most prominently in the researched setting. These topics are the related issues of responsibility and accountability. While we see these two issues as very closely related to transparency and liability as discussed in the first case, they highlight subtly different aspects. For example, accountability focuses on the trading of accounts between actors, while transparency highlights how previously non-visible actions or assets gain visibility within and outside of the organization. In the luxury goods industry, the technologies were being discussed within the context of industry-specific risks such as counterfeits and an increasing interest in being able to combat the flow and use of fake products. Technological solutions in the form of ubicomp type technologies were seen as a countermeasure: They would allow unique identification of each product, could provide track and trace data and a product verification infrastructure to identify risky products and additionally could be used as safety features on products to make tampering more difficult (e.g., by attaching RFID chips). Designers, trial users, and industry partners agreed that from a security point of view, it would be advantageous to use such a system to combat counterfeits. In our study, we aimed to show the complexity of the two topics of responsibility and accountability within organizational settings and uncovered how they looked different from different perspectives which could lead to actors faced with the technologies finding themselves in contradictory situations requiring that (ethical) dilemmas be solved in order to be able to accomplish the work.

2.2.1 Responsibility and Accountability

We encountered one example for a dilemma pertaining to responsibility and accountability when talking to people in charge of shops. The example concerned the idea of a shop employee verifying a product while serving a customer. For example, a readout of the data stored on an RFID chip attached to a watch might be performed when the watch is sold or when a product is brought back for servicing. However, it was pointed out that such a check could constitute a problem as any use of an RFID system at the point of sale visible to the customer might interfere with the client relationship and the romance of the environment and might invoke a trust issue for the shop. In addition, a visible check in a repair situation could indicate a lack of trust toward the customer. Also, an indication that the watch might be a counterfeit could change the behavior of the shop employee and could be irritating for the customer. Within the research project, it was therefore discussed how to hide the check from the customer. The project partners feared a lack of control over the situation if there was the possibility of an unexpected alert in a sales situation. The new responsibility to verify luxury goods in front of customers conflicted with the existing responsibility to always provide a pleasurable environment and to offer only genuine products to the customer. Also, it was argued that employees did not want to be held responsible for a problem with the product originating in an external source.

The problems explained here could be solved by the shop employees performing the check out of sight of the customers. However, any solution involves a trade-off between two conflicting interests: how to guarantee the genuineness of luxury goods and the varying interests of stakeholders where authentication needs to be performed.

The shop employees' responsibility to provide a safe and pleasurable environment in the local situation of sale or support clashes with the more general responsibility of the industry to guarantee the high quality and genuineness of the products. The latter responsibility led to a technology system becoming attractive which would enforce "responsible" behavior—while at the same time reducing a local actor's capabilities for controlling a typical sale situation. From this example we realized that there could be misfits between multiple responsibilities.

The example also showed us that accountabilities and control capabilities were crucial elements in ubicomp technologies' role in ethical dilemmas. Ubiquitous computing technologies such as the system for tracking luxury goods are discussed to provide accounts to be able to satisfy accountability demands. However, as we can see from our example, there can be multiple and even conflicting accountabilities that come into play. The sales person is accountable to the customer as well as to the verification system and struggles to satisfy both demands. This becomes especially problematic if he or she is not given the necessary capabilities of control to satisfy the demands for accounts, for example, if she sees an alert and is asked to react in time but actually is not able to do so, because she would break the romance of the sales situation [32].

By introducing a ubicomp system, the luxury goods industry was aiming to improve the trustworthiness of products. However, they were also—and so were the individual employers and employees—participating in defining what counts as trustworthy in our contemporary societies. A specific way of providing accounts, namely by relying on data collected with a ubiquitous computing system, was gaining importance. While an increasing demand for accountability is part of a larger societal trend [33] such demands become particularly prevalent in workplaces when new demands posed by and with IT technologies become entangled with already existing accountability demands [34, 35]. The distributed and networked nature of ubicomp applications often makes it difficult to predict, if and in what way the new systems can increase transparency and answer demands for accountability. Conflicting accountability demands need to be central in both systems designers' attention and also in the attention of those making decisions about rolling out ubicomp systems. Complex and connected settings that are the target of ubicomp always also include individual workplaces where accountability conflicts can lead to dilemmas hard to solve for individual actors.

To summarize, it became clear in the case of tracing luxury goods that ubicomp technologies led to new accountability challenges as control shifts can occur with the new availability of data in places where this data had not previously been available to actors.

3 The Bigger Picture: What Societal Trends?

With regard to our central aim in this paper, namely to map the bigger changes that are part of the causes for the new ethical challenges that people face in workplaces, we can summarize as follows: Companies and even whole industries are seen to be under pressure to become transparent and to provide proof and accounts of everyday work activities. Information technology in the form of ubicomp "...generates information about the underlying productive and administrative processes through which an organization accomplishes its work. It provides a deeper level of transparency to activities that had been either partially or completely opaque" [36, p. 9]. Accounts facilitated by information technology are increasingly important in order to satisfy transparency demands [37]. Increasingly, we expect retrievable, reliable, high-quality data about everything that occurs and rely on it to make decisions, to verify, or to blame. Through our expectations, more and more areas become available for data capture through sensors, new recording, and satellite technologies. Rather than ubicomp itself being the significant novelty, it is this trend toward *informational ubiquity*, which enables the rising importance of providing proof and becoming transparent.

With accounts becoming more and more important, there is also an increasing attention to being able to prove what has been done, when, by whom, and how, rather than it being important whether a task is actually performed. Strathern has described this phenomenon as an audit culture, in which we are increasingly concerned with being expected to provide verification to satisfy the demands of accountability. As an instrument of accountability, audit culture is hard to criticize as it advances values such as openness, responsibility, and widening of access.

In "rituals of verification" management practices originating from protocols of financial accountability have become a "taken-for-granted process of neo-liberal government and contributing substantially to its ethos" [33, p. 3]. The rise of an audit culture is linked to the emergence of technologies that facilitate the monitoring of previously inaccessible domains and enable the new regime of audit.

4 Ethical Challenges in Introducing Ubicomp

With regard to the ethical challenges faced by those introducing ubicomp in the workplace and by those who use it in their workplaces, we have attempted to show how larger trends such as an increasing attention toward audit, transparency, and control over complexity play out in concrete workplace settings. Gauging whether the introduction of a technology will be beneficial or detrimental to certain actors and their interests is very difficult due to the complexity and interconnectedness that are integral elements of an IoT system. In addition, individuals are faced with dilemmas pertaining to making decisions based on conflicting accountability and responsibility demands. Giving the various individuals within an organization the necessary control over both the system and the information it creates—in order to allow individual decision making and dealing with idiosyncratic and context-dependent ethical, and other dilemmas—is a complex and difficult challenge that organizations face.

From our case studies, we suggest that—in addition to considering "privacy" or "surveillance" which often seem to be used as black boxes for the issues of control over information explored here—decision takers within organizations and those regulating the field of intelligent working environments should take into account:

- what may be the results of an increased transparency of actions and asset properties that the new technologies provide, especially with regard to the viewpoints of individual employees;
- how liability issues come into play in creating external pressure to introduce such technologies and also in changing ideas within the organization about what type of information counts as reliable;
- how responsibilities may shift between individual employees, the new system, and external bodies (such as a regulatory board for health and safety or an industry-wide initiative to prevent fraud);
- and how to ensure that actors' control capabilities are sufficient to satisfy newly arising accountability demands.

The ethical challenges that those face who are tasked with deciding whether to introduce a ubicomp technology and also the ethical dilemmas that can occur for those who use such technologies are connected to wider societal trends of informational ubiquity. The visibility of processes and the transparency of product chains and work flows are not only affordances of new technological developments, but we find ourselves in socio-technically produced audit cultures in which control over information is all-important. While in individual workplaces issues can be mitigated by ensuring sufficient control capacities in the concerned individuals, this is not so easy for managers having to consider the bigger picture of their organization within the overall industry. The pressure to compete and to comply may leave no choice but to adopt certain technologies, but managers can aim to ensure that benefits and drawbacks are equally distributed within the organization without a one-dimensional empowerment of only one group.

Acknowledgments   The initial field studies that this work is based on were supported by the UK Engineering and Physical Sciences Research Council (EPSRC) project NEMO (EP/C014677/1) and by the Swiss ETH Research Grant TH-31/06-1. We would like to thank the EU funded project Stop Tampering of Products (SToP) (IST-034144) for sharing of insights.

References

1. Orlikowski, W.J.: Sociomaterial practices: exploring technology at work. Organ. Stud. **28**(9), 1435–1448 (2007)
2. Weber, R.: Internet of things—new security and privacy challenges. Comput. Law Secur. Rev. **26**(1), 23–30 (2010)
3. Ehrwein Nihan, C.: Healthier? more efficient? fairer? an overview of the main ethical issues raised by the use of ubicomp in the workplace. ADCAIJ: Adv. Distrib. Comput. Artif. Intell. J. **1**(4), 29–40 (2013)
4. Philips, D., Wiegerling, K.: Introduction to IRIE. Int. Rev. Infor. Ethics **8**, 4–6 (2007)
5. Bohn, J., Coroamă, V., Langheinrich, M., Mattern, F., Rohs, M.: Living in a world of smart everyday objects—social, economic, and ethical implications. Hum. Ecol. Risk Assess. **10**(5), 763–785 (2004)
6. Kinder-Kurlanda, K.E., Ehrwein Nihan, C.: Ethically Intelligent? a framework for exploring human resource management challenges of intelligent working environments. In: van Berlo, A., Hallenborg, K., Corchado R., Juan M., Tapia, D.I., Novais, P. (eds.) Ambient Intelligence —Software and Applications, 3rd International Symposium on Ambient Intelligence (ISAmI 2012), pp. 213–219. Advances in Intelligent Systems and Computing, 219, Springer, Berlin (2013)
7. Van Kranenburg, R., Bassi, A.: IoT challenges. Commun. Mob. Comput. **1**(1), 1–5 (2012)
8. Latour, B.: Reassembling the Social: An Introduction to Actor-Network-Theory. Oxford University Press, Oxford (2005)
9. Zammuto, R.F., Griffith, T.L., Majchrzak, A., Dougherty, D.J., Faraj, S.: Information technology and the changing fabric of organization. Organ. Sci. **18**(5), 749–762 (2007)
10. Orlikowski, W.J., Scott, S.V.: Chapter 10 sociomateriality: challenging the separation of technology, work and organisation. Acad. Manage. Ann. **2**, 433–474 (2008)
11. Leonardi, P.M., Barley, S.R.: What's under construction here? social action, materiality, and power in constructivist studies of technology and organizing. Acad. Manage. Ann. **4**(1), 1–51 (2010)
12. Suchman, L.: Plans and Situated Actions: The Problem of Human-Machine Communication. Cambridge University Press, Cambridge, New York (1987)
13. Bijker, W.E., Law, J.: Shaping Technology/Building Society: Studies in Sociotechnical Change. Institute of Technology, Massachusetts (1994)
14. Bijker, W.E., Hughes, T.P., Pinch, T.: The Social Construction of Technological Systems: New Directions in the Sociology and History of Technology. Institute of Technology, Massachusetts (2012)

15. Clegg, C.W.: Sociotechnical principles for systems design. Appl. Ergon. **31**, 463–477 (2000)
16. Kallinikos, J.: The Consequences of Information: Institutional Implications of Technological Change. Edward Elgar, Cheltenham (2006)
17. Pasquinelli, M.: Die Regierung des digitalen Mehrwerts: Von der NetzGesellschaft zur Gesellschaft der Metadaten, In: Kulturaustausch, 3 (2010), e-Volution, Institut für Auslandsbeziehungen, Berlin (2010)
18. Schoen, H., Gayo-Avello, D., Metaxas, P.T., Mustafaraj, E., Strohmaier, M., Gloor, P.: The power of prediction with social media. Internet Res. **23**(5), 528–543 (2013)
19. Brynjolfsson, E., McAfee, A.: The Second Machine Age: Work Progress, and Prosperity in a Time of Brilliant Technologies. W. W. Norton & Company, New York (2014)
20. Levy, F., Murnane, R.J.: The New Division of Labor: How Computers are Creating the Next Job Market. Princeton University Press, Princeton and Oxford (2012)
21. Durkheim, E.: The Division of Labour in Society. Palgrave Macmillan, London (1984)
22. Hannerz, U.: Cultural Complexity. Columbia University Press, Studies in the Social Organization of Meaning. New York (1992)
23. Kinder, K.E., Ball, L.J., Busby, J.S.: Ubiquitous technologies, cultural logics and paternalism in industrial workplaces. Poiesis Praxis: Int. J. Technol. Assess. Ethics Sci. **5**(3–4), 265–290 (2008)
24. Boos, D., Grote, G.: Designing controllable accountability of future internet of things applications. Scandinavian J. Inf. Syst. **24**(1), 3–28 (2012)
25. Boos, D., Guenter, H., Grote, G., Kinder, K.: Controllable accountabilities: the internet of things and its challenges for organisations. Behav. Inf. Technol. **32**(5), 449–467 (2013)
26. Fleisch, E., Tellkamp, C.: The business value of ubiquitous computing technologies. In: G. Roussos (eds.) Ubiquitous and Pervasive Commerce, pp. 93–113. New Frontiers for Electronic Business, Springer, London (2006)
27. Lee, H., Özer, Ö.: Unlocking the value of RFID. Prod. Oper. Manage. **16**(1), 40–64 (2007)
28. Strassner, M., Schoch, T.: Today's impact of ubiquitous computing on business processes. In: First International Conference on Pervasive Computing, pp. 62–74, Short Paper Proceedings, Zurich, August 2002
29. Lehmann, R.J., Reiche, R., Schiefer, G.: Future internet and the agri-food sector: state-of-the-art in literature and research. Comput. Electron. Agric. **89**, 158–174 (2012)
30. Fano, A., Gershman, A.: The future of business services in the age of ubiquitous computing. Commun. ACM **45**(12) (2002)
31. Kinder, K.E.: Ubiquitous Computing for Industrial Workplaces: Cultural Logics in Designing for the 'Real World'. Intelligent Environments, IET International Conference on Intelligent Environments (Seattle), 21–22, 1–5 (2008)
32. Lehtonen, M., Boos, D., Graf von Reischach, F., Magerkurth C., Müller, J., Bogataj, K., Gout, E., Gourmanel, F., Ippisch, T., Oertel, N., Dada, A.: Stop Tampering of Products, Deliverable 1.5, Final Evaluation of Project Results According to the Requirements Identified. Karlsruhe, SToP (2009)
33. Strathern, M.: Introduction: New accountabilities. In: Strathern, M. (ed.) Audit Cultures: Anthropological Studies in Accountability, Ethics and the Academy, pp. 1–18. Routledge, London, New York (2000)
34. Wintereik, B.R., Van Der Ploeg, I., Berg, M.: The electronic patient record as a meaningful audit tool: accountability and autonomy in general practitioner work. Sci. Technol. Human Values **32**(1), 6–25 (2007)
35. Yakel, E.: The social construction of accountability: radiologists and their record-keeping. Inf. Soc. **17**(4), 233–245 (2001)
36. Zuboff, S.: In the Age of the Smart Machine: The Future of Work and Power. Basic Books, New York (1988)
37. Neyland, D., Woolgar, S.: Accountability in action? The case of a database purchasing decision. British J. Sociol. **53**(2), 259–274 (2002)

Ubiquitous Computing in the Workplace: Ethical Issues Identified by the Interdisciplinary IWE and HRM Research Group

Céline Ehrwein Nihan

Abstract This article presents a synthesis of the first results of the intelligent working environment and human resources management (IWE and HRM) project. I will start with some considerations aimed at *clarifying the conceptual and general epistemological frameworks* which underline this research (1). I will then turn to the axiological and normative context which surrounds the development of IWEs. In a *descriptive* approach, I will draw up the main moral values at stake identified by our research group and will try to clarify their meaning (2). The third section will be devoted to the ethical *challenges* raised by the emergence of the IWEs (3). These will be considered from two different and complementary perspectives. The first one, which will follow a *deontological* approach, will lead me to highlight the conflicts that might arise between the principles at stake, as well as between the different normative systems supported by the stakeholders. The second perspective, which will be more *consequentialist*, will bring me to describe some possible positive and negative impacts of the development of the IWE in the eyes of both employers and employees. I shall conclude with a few points of reflection on the ethical rules that should be respected in the development and implementation of Ubicomp in the workplace (4).

Keywords Ethics · Ubiquitous computing · Workplace · Intelligent working environment

C. Ehrwein Nihan (✉)
University of Applied Sciences in Business and Engineering Vaud (HEIG-VD),
Av. des Sports 20, 1400 Yverdon-Les-Bains, Switzerland
e-mail: celine.ehrwein@heig-vd.ch

© Springer International Publishing Switzerland 2015
K. Kinder-Kurlanda and C. Ehrwein Nihan (eds.), *Ubiquitous Computing in the Workplace*, Advances in Intelligent Systems and Computing 333,
DOI 10.1007/978-3-319-13452-9_6

1 Conceptual and Epistemological Framework[1]

One of the first questions that needed to be addressed was the question of the normative and axiological context in which the development of IWEs takes place. As we shall see below, we had to identify the moral principles[2] involved and try to clarify, as far as possible, their import and meaning. However, it quickly became apparent that this question presupposed fundamental conceptual and epistemological issues. I would like to briefly examine these issues here.

1.1 IWEs: Reality or Utopia?

First issue: the reality of IWEs.

During the early stages of our project, we often heard—from HR managers in particular—that our topic was irrelevant since IWEs did not exist or were not, or at least not yet, implemented in companies and that, consequently, our research was unfounded. While the literature especially from computer science paints a very different picture, in this context of doubts and questioning, it seems important to stress the reality of IWEs. These are neither a utopia nor a futuristic dream: They are a very real and clearly identifiable phenomenon.[3]

To ensure that in our reasoning, we stayed close to actual developments, our group examined a series of articles on various aspects of information and communication technology (ICT) research. We paid particular attention to applications developed for the world of work. On this basis, we developed the following definition, which formed the basis of all our discussions:

> IWEs are working environments fitted—often imperceptibly—with one or more ubiquitous computing systems which record, integrate, correlate, and analyze ambient data from diverse sources and are intended to meet the needs

---

[1] Regarding the methodological aspects of the IWE and HRM project, see also [1].

[2] Contrary to the traditional use in the ethics literature, the terms "value" and "principle" are used here interchangeably in order to designate a good to which we attribute an axiological (orientation of the action) and normative (limitation of the action) force.

[3] In fact, these doubts regarding the reality of IWE belong to a broader phenomenon which concerns the whole of Ubicomp. Indeed, as Dourish and Bell have shown [2], the vision proposed by Weiser in the late 80 carries with it the idea of an unreachable future. Now, this idea prevents us from seeing the reality that surrounds us: "Ubicomp's proximate future continually places its achievement out of reach, while at the same time blinding us to current practice" (22). Of course, ubicomp has not yet deployed its full potential. But it is not just a hypothetical future: it has already begun to fundamentally transform our lives. "The future is already here. The technological trends that Weiser insight fully extrapolated have resulted in radical transformations and reconfiguration in everyday life, just as he anticipated" (41).

of the stakeholders automatically, in due time and in a personalized and intelligent manner.

That being said, we must also recognize—and this is a constant element in the research—that there is a gap between the technological advances, their placing on the market, and the public awareness of their reality and of the possibilities they offer. Today, neither managers nor employees seem to be sufficiently aware of the radical changes that Ubicomp brings regarding the organization of work and professional relations.[4] This is the reason why we thought this was a subject worth studying.

1.2 What Comes First? IWE and Moral Values: A Reciprocal and Dynamical Relation

Second issue: the relationship between the axionormative social framework and technological developments.

The two anthropologists who took part in our project drew our attention to the *reciprocal and dynamic nature* of these relations. In other words, they insisted on the importance of rejecting the erroneous idea that technological developments *unilaterally* influence our systems of values or normative principles. Of course, the deployment of Ubicomp questions the existing moral framework. They affect the meaning and relevance of our common shared values, and contribute thereby to their evolution. They can lead to the emergence of new normative sets and new axiological balances. But technological developments are also influenced in turn by the moral context in which they take place. To say it in other words, IWEs are not ex nihilo creations, springing from nowhere. They take place within an axionormative framework which they partially reflect.

1.3 Who Is (Primarily) Concerned? Identification of the Stakeholders and Clarification of Their Status

Third issue: the identification of the stakeholders.

If we are to consider in the most complete and most possible in-depth manner, the ethical issues raised by the development of the IWEs, we also have to identify in advance the persons concerned and must try to clarify their status. That is what we did (see Table 1). By doing so, we were led to distinguish between the persons

---

[4] See in particular [3].

Table 1 Stakeholders
involved in the development
of IWE

Directly concerned stakeholders
Employers (in particular CIO and HR Managers)
Employees
Indirectly concerned stakeholders
Designers
Producers
Clients/service recipients
State
Insurances
Civil society
Unions

Sources Synthesis of the first results of the IWE and HRM project

directly concerned by the deployment of Ubicomp in the workplace (that is the "users" of IWEs) and those who are more *indirectly involved* (their developers and those who may be affected or may benefit from them without necessarily be seen as first users of the system).

In the context of our project, we chose to focus on the issues raised by IWEs for those who are a priori more directly concerned by their development, namely (HR) managers, on the one hand, and employees, on the other hand. The other stakeholders were considered only insofar as their positioning and behavior could have an influence on the mainly involved groups.

1.4 Assumptions Regarding the Attitude of the Stakeholders Toward IWE

Fourth issue: the place and role of stakeholder-specific systems of values regarding their attitude toward IWEs.

In our discussions, we postulated that the attitude of the various stakeholders toward the IWEs depends on their specific system of values, but also on other factors such as their position in the organization, their experience and personal backgrounds, or their knowledge of the Ubicomp technology. Overall, we worked on the basis of the following assumptions.[5]

1. The (more or less open, supportive and suspicious, etc.) *attitude* or *degree of acceptance* toward IWEs of the different stakeholders depends at least in parts

---

[5] These assumptions should be checked at a later stage of our project.

on their appreciation of the shared values and principles that accompany the development of Ubicomp, particularly in the workplace.

In other words, a person (or a group of people) who thinks that his or her autonomy is more important than his/her health will a priori not have the same attitude toward an intelligent system monitoring his/her fatigue than a person who on the contrary considers that health prevails over self-government.

2. *The assessment of the values and principles surrounding the development of the IWEs* by the different stakeholders is in part dependent on their understanding of these environments, their (personal, organizational, social, etc.) interests and their (financial, intellectual, practices, managerial, etc.) goals.

We may reasonably presume that an insurer whose mission is to reduce the costs generated by work accidents will probably put more emphasis on control as an essential professional value than the average employee who does not run special risks in the execution of his job.

3. Each stakeholder has a particular *understanding* of IWE and is driven by his/her *own interests* and *goals*—which are more or less identifiable, but also changeable.

It seems obvious that a designer and/or an engineer specialized in ICT does not have a priori the same knowledge of the potential of IWEs as a mere employee, who has no special expertise in IT. Similarly, these two types of people are likely to have partly different interests and goals (personal and professional) to defend.

4. The (personal and cultural) background of the various stakeholders as well as *their position toward and/or within the organization* has an impact on their understanding of the IWEs, their interests, and the objectives they pursue.

We can quite easily assume that belonging to a particular social or professional group (employees, employers, designers, trade unions, etc.) as well as the positioning of this group toward the global functioning of the company (internal/external stakeholder, contractual non-contractual relations, position at the top/bottom of the hierarchy, etc.) has an impact on the (technical, commercial, etc.) understanding that the different stakeholders may have of IWEs, as well as on their interests and their objectives. It appears most likely that the expectations that are placed on the manager for collective performance will, for instance, bring differences in perception compared to those of subordinate employees regarding the system's ability to detect a drop in productivity. Similarly, even in the case where an employer and an employee are both globally favorable to the development of IWEs, they will probably not defend these for the same reasons (Fig. 1).

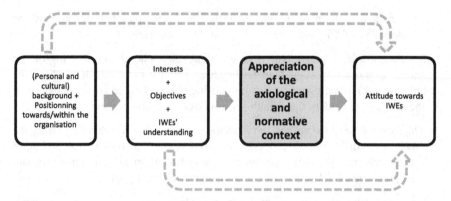

Fig. 1 Place and role of the axiological and normative context regarding the attitude toward IWEs

1.5 Objectives of the Research Project

With our project we want to:

- Identify the implicit and explicit axiological and normative context which surrounds the development of IWEs. In other words, we seek to highlight both the told and the untold stories of the key values and principles at stake and the major trends which characterize their evolution.
- Analyze these elements from the perspective of the different stakeholders, that is to say, with the assumption that each group of people involved possesses a specific understanding of IWEs and pursues, whether consciously or not, its own interests and goals.

In doing so, our objective is to:

- Identify the main ethical challenges raised by the deployment of Ubicomp in the workplace. This objective is consistent with a central aspect of the precautionary principle since we try to anticipate the potential consequences of the growing implementation of IWEs and to consider what measures could be taken to avoid problems or ethical conflicts.
- Provide tools intended to help key stakeholders to cope with these new challenges.

2 Axiological and Normative Common Context

The development of IWEs is part of a context which consists of both (moral or legal) norms and (individual, interpersonal, and social) values.

In our research, we are particularly interested in the *axiological and normative* aspects which underlie the current technological developments. We focused our attention on the values and principles that are put forward in order to explain, justify, and/or challenge the merits of the development of Ubicomp in the workplace.

The list presented below is not exhaustive. Nevertheless, it reflects the values and principles which are most often mentioned in the literature related to IWEs. That is the reason why we think that employees and employers will give priority to them in order to provide a basis for their attitude toward the developments of Ubicomp in the workplace.[6]

Privacy

The most often and most explicitly mentioned value is that of privacy (see for instance [4–7]). The importance given to it in the field of computer ethics is not new [8]. But with the development of ambient intelligence, our understanding of the signification and the limits of privacy seems particularly questioned.

According to Floridi [9], ambient intelligence even changes its nature. As noted by Heesen and Siemoneit [10]:

> Three different forms of privacy are commonly distinguished (a) Decisional privacy which refers to the level of freedom of decision. (b) Local privacy which has to do with the protection of living quarters and of residence information but also with the safeguarding of corporal integrity. (c) Informational privacy which describes the protection and control of person-related information.

Ubicomp deals with these three areas. Its sensors make possible the measurement and processing of a growing amount of personal (emotional, physiological, etc.) data. Increasingly becoming autonomous, it is able to manage in our place an ever greater number of actions and decisions. Finally, ubiquitous computing has for characteristic that it can pry into different social spaces (home, work, streets, etc.) without being noticed [8, 11].

Autonomy

Closely related to privacy, autonomy is also often mentioned when it comes to justifying or challenging the merits of Ubicomp [7, 10]. Assuming the characteristics of a real agent [9, 12, 13], IWEs seem to be able both to increase and to limit human skills and capabilities.

> Ambient intelligence and persuasive technology have an ambivalent relationship toward human freedom. Whereas in many cases they have been designed to create freedom, as they quietly relieve us of all sorts of tasks, they also form a threat to this very freedom, because they influence and control us [14, p. 236]

Health

Health appears very frequently in discussions related to ambient intelligence [13, 15–18]. With their multiple sensors, IWEs seem to be able to perform a full physical and psychic monitoring (heart rate, breathing, emotions, fatigue, etc.) of the persons who find themselves surrounded by them.

---

[6] Obviously, these values play a role beyond this one area: they apply globally to recent developments in ICT and, more generally, to all fields of activity in which we are engaged.

Safety
Safety is usually mentioned along with health [15, 16]. In fact, Ubicomp is often presented as a technology which is useful for the prevention of unexpected events (accidents linked to inattention, air pollution, etc.).

Security
Security is also regularly invoked in the debate on ambient intelligence. Most of the time, it appears in connection with the issue of privacy (protection against data falsification, hacking or misuse of personal data, etc. (see e.g. [17]). As security, it refers to a certain ideal of mastery and control.

Control
The importance of this principle manifests itself essentially in three ways. Several authors highlight fears of a *loss of control of humans on technological tools* [19, 20]. In this case, control is usually closely associated with the protection of autonomy and self-government. As noted by Boos et al. [21], it refers then to the requirements of transparency, predictability, and influence.

But reference is also often made to the capacity conferred by ambient intelligence to *keep a watch over or on the individuals*. The question then arises whether and to what extent the control which is made by the machine impact on our behavior and on power relations [11, 15, 16, 22, 23].

Finally, ambient intelligence is often presented as a tool which helps us to deal with particularly complex situations. With this in mind some authors question the *ideal of mastering reality* carried on by the Ubicomp and its potential dangers [24, 25].

Responsibility
Safety, security, and control all refer to the issue of liability. In the literature, this issue is considered from very different perspectives which are globally built around three types of questions.

The first concerns the *degree of responsibility and accountability*. Thus, as noted for example by Verbeek [14], "The question arises of whether we can still be held entirely responsible for actions induced by these technologies." For some authors, the issue goes even further, since, according to them, we need to raise the question of the moral/legal status of virtual agents [9, 12].

The second question concerns the *content of this responsibility*. It is about asking precisely what, on which level, and to which extent the various stakeholders are responsible (for the design of Ubicomp system, its potential use/misuse, etc.).

Finally, some scholars question the *conditions of possibility of responsibility* at the age of ambient intelligence [19, 21] and try to sketch its (new) requisites.

Justice
Although they are less often and less explicitly mentioned, the notions of justice, fairness, and equality (between people and institutions) also appear in the Ubicomp discussion [8, 13]. Ambient intelligence is thus presented as encouraging discriminations as well as being able to fight against (natural, social, cultural, etc.) disparities.

Performance
Performance occupies a relatively important place in the discussion IWEs [3, 26]. It is generally associated with the values of control, autonomy, health, and safety—the idea being that the empowerment of personal skills and the reduction in risks (diseases and accidents), which are both possible thanks to the support of the machine, should infer positive effects on the individual and collective performances.

Social interactions and integration
The "fathers" of ambient intelligence "wanted to put computing in its place, to reposition it into the environmental background, to concentrate on *human-to-human* interfaces and less on *human-to-computer* ones" [27]. Today, many authors continue to emphasize the socializing power of ambient intelligence [28]. However, this view is not unanimous [29–31].

3 Ethical Challenges

3.1 Deontological Approach[7]

In our opinion, the values or principles mentioned above constitute the common axiological and normative framework to which employees and employers will refer in order to base their attitude toward IWEs. In other words, a priori, these are *shared moral values and principles*.

This being said, having in common moral principles is not a sufficient condition to prevent ethical conflicts.

3.1.1 Main Deontological Conflicts Within the Axiological and Normative Common Context

The first kind of conflict that may arise in relation to the development of IWEs lies in the moral principles themselves. Indeed, these principles are likely to conflict, irrespective of the particular context in which they apply.

Table 2 sums up the main axionormative conflicts which, according to us, may appear in the context of the development of the IWEs.

---

[7] In ethics, it is usual to make a distinction between what we call deontological and teleological approaches. While the first focus on the moral *imperative* or rules which have to be respected regardless of any other considerations (the term "deontological" comes from the Greek *deon* = duty), the second imply on the contrary to give special attention to the *objectives* of our deeds (the term "teleological" originates from the Greek *telos* = goal). Consequentialism is one particular form of the teleological approach which implies to take into consideration the potential concrete *impacts* (or consequences) of our actions and decisions.

Table 2 Main value conflicts

Control (surveillance of people)	↔	Privacy
Health		Performance
Autonomy		Control (surveillance of people)
		Health
		Safety
		Social interactions and integration
		Autonomy

Sources Synthesis of the first results of the IWE and HRM project

Thus, it is very often noted that the ideal of people's *surveillance* conveyed in particular through IWEs may undermine the respect for *privacy* [4, 11]. That said, conversely, it can also be argued that, in certain situations, an excessive appreciation of the right to privacy is contrary to the legitimate expectation of (social and organizational) control over behaviors that are seen as problematic or dangerous.

A conflict may also arise between the ideal *performance* at work, on the one hand, and the value of *health*, on the other hand. Indeed, encouraging employees or managers by the means of IWEs to be more efficient could in the long run cause stress with adverse health consequences [3].

That said, tensions are mainly concentrated around the principle of autonomy.

First, the ideal of autonomy is likely to run counter to the requirement of *people's surveillance*. One of the main issues that arises here is how someone who knows that he/she is potentially observed is able to maintain his/her freedom of action and behavior [32].

In addition, autonomy is particularly inclined to conflict with *health* and *safety*. Indeed, as Kinder et al. [16] point out, a boss who imposes rules of conduct to subordinates on the pretext that it is for their own good is a traditional feature of working relationships which ambient intelligence then reinforces.

Finally, depending on the circumstances, autonomy could thwart our expectations for *integration and social interactions*. By promoting employees' empowerment, ambient intelligence may indeed potentially pushing them to retreat into themselves and to move away from the others.

We should add here one more very special axiological conflict, that is, the conflict which may *oppose autonomy to autonomy*, or to be more precise, the autonomy of a person or of a group of persons (for instance the employees) to the autonomy of the others (for instance the managers). Now, since "having autonomy" means "having power over," the opposition between autonomy requirements implies a radical questioning of power relations.

Now, this conflict plays a fundamental role in the development of IWEs. Moreover, if we assume, as it is often the case today, that "data is the new oil"—in the sense that it gives power to the person who possesses it, then it seems possible to say that people (employees, employers, engineers, designers, etc.) and/or institutions (business, insurance, government, unions, etc.) that control the largest amount of data have a priori more power than others.

However, one of the special features of ambient intelligence is that, more than any other technology before, it offers the opportunity to collect and handle in real time an impressive amount of information and this, potentially, without the knowledge of its "users." To put it in other words, Ubicomp seems to be able to question, through a more or less imperceptible process, the very balance of (interpersonal, institutional, and social) power relations [16, 22].

3.1.2 Conflict Within the Different Stakeholders' Axiological and Normative Systems

The of a *conflict between the various axionormative systems* should be added to the risk of a conflict between values. Indeed, each shared value has a wide semantic field: Thus, its impact and meaning can vary depending on the people who relate to it, how the people relate to it, and on the particular contexts of action to which it applies.

So, as I said, the position a person occupies in the hierarchy may have a certain impact on the way he/she will appreciate the shared moral system. To put it differently, it is likely that employers and (HR) managers will not give the exact same weight and significance to the values that compose their common axionormative framework.

Thus, we assume (though this will need to be checked) that a manager who exercises responsibility will more easily tend to bring forward the performance and surveillance aspects than an ordinary employee. In contrast, employees in a subordinate position will probably be more sensitive to the issues of privacy and autonomy at work than their superiors.

Table 3 tries to imagine which values will a priori be more strongly put forward by employees and which ones will rather be highlighted by employers.

Table 3 Appreciation of values by the stakeholders

Values which are mostly put forward by employees	Privacy Autonomy Health Control (over the system) Justice (do not suffer <u>personal</u> discriminations other than positive ones) Social interactions and integration
Values which are mostly put forward by (HR) managers	Control (surveillance of people and mastering reality) Justice (do not suffer <u>institutional</u> discriminations other than positive ones) Safety Security Responsibility Performance

Sources Synthesis of the first results of the IWE and HRM project

If our hypothesis is correct, it is likely that differences in axionormative assessments will lead to significant disagreements between employees and (HR) managers, which could lead to conflicts regarding the implementation and management of the IWEs.

3.2 Consequentialist Approach

The analysis of the conflicts that may occur with respect to the axionormative framework is only one possible approach to the ethical challenges raised by the development of IWEs. This approach, which can be described as deontological, is important insofar as it allows us to put current developments in their broader social context and to appreciate them under the perspective of our common shared values and principles.

However, in our opinion, this approach deserves to be extended and complemented by a questioning of the (positive or negative) *consequences* that occur or may occur with the development of Ubicomp in the workplace.

In order to provide a clear frame for these questions, we chose to base our presentation on the values that had been identified and, from there, to briefly describe the main (real or expected) effects of ambient intelligence in the workplace.

Privacy	
+	IWEs permit the anonymization of some work processes [9]. Thus, they provide, in principle and at least in some cases, a better protection of privacy. With ubiquitous computing system, a manager may, for instance, more easily evaluate the work of his team without necessarily having to know the details of each individual performance.
−	The fact that ambient intelligence is a priori imperceptible may encourage spying on employees by employers. IWEs can detect health problems that employees may prefer not to know about (see the discussions in bioethics on this topic; see also [33]).
Autonomy	
+	Thanks to the IWEs' support, workers may in some cases develop new skills and so increase their professional autonomy.
−	By getting more autonomy, workers may become less dependent on the presence and help of their colleagues. They may be tempted to distance themselves from others or indeed to give up solidarity they have with their colleagues. IWE may also change the balance of power within a company and thus promote inequalities.
Health	
+	The collect and process of physiological data (heart rate, breathing, etc.) through IWEs permit to check employees' health (fatigue, stress, etc.). Thus, the latter may benefit from advice to keep in good shape and to reduce their risk of becoming ill. Since they provide better health monitoring, IWEs should have positive effects on the corporate performance [3].

(continued)

(continued)

Health	
−	The implementation of IWEs induces changes in professional roles and practices [3]. The uncertainties related to these changes may cause additional stress for employees that may affect their health. As Hilty et al. have noted [29, 34], we cannot exclude that the use of ambient intelligence may have direct effects on employees' health and favor certain diseases (cancer, sleep disorders, etc.). Finally, the argument that IWEs may provide benefits to occupational health may be used to support paternalistic attitudes of some managers and lead to a denial of the right of subordinates' self-determination [22].

Safety	
+	Better than any other technological tool so far, IWEs permit users to anticipate, detect, and prevent various accidents related to employees' inattention, stress, or fatigue, as well as to the use of chemicals or of industrial machines.
−	Like the health argument, the safety argument may support paternalistic logic underlying the world of work and lead to a denial of employees' autonomy. Paradoxically, the arguments that promote the strengths of IWEs in work safety may cause a greater sense of insecurity [35].

Security	
+	IWEs, as indeed ICTs in general, allow a processing and storage of data that, in some respect, ensure better protection against malevolent acts (theft or destruction of data) than traditional tools [9].
−	In the same way, as for safety issues, arguments that highlight the contribution of IWEs in terms of protection against malevolent acts may, paradoxically, cause a greater sense of insecurity [35]. Programming errors, virus, or hacking may lead people (employees, employers, or third parties) to process erroneous data without realizing it.

Control over the system	
+	/
−	Delegating complex tasks to intelligent systems may lead to: • a loss of workers' control over the machine; • a loss of some human and professional skills; • a loss or a dilution of responsibilities [29].

Surveillance of people	
+	A priori ambient intelligence provides employers with the opportunity to better control (greater objectivity, more regular monitoring, etc.) workers' activity and to react more quickly to their potential errors or loss in productivity.
−	As mentioned, the use of ambient intelligence to monitor employees' activity may harm their privacy and cause additional stress among workers. The knowledge of being potentially monitored by IWEs may affect workers' spontaneity and get them to behave in a standardized way.

Mastering reality	
+	IWEs may take on the management of complex professional activities. In doing so, they are expected to: • lighten employees' work and contribute to decrease their stress; • reduce the risk of professional errors and accidents; • increase the corporate performance and competitiveness.

(continued)

(continued)

Mastering reality	
−	Presented as fully safe and efficient systems (especially for the treatment of complex tasks), IWEs may induce a blind trust in technology and the illusion of omniscience among users (employees, managers, etc.).

Responsibility	
+	The monitoring of professional activities by IWEs should facilitate the assignment of responsibilities (in companies or insurances), especially when errors are made by an employee.
−	We can expect that increased workplace monitoring will reinforce managers' expectations toward workers. By putting more and more pressure on employees to assume their responsibilities, IWEs could have negative effects on their health. Audit culture can lead to providing proof of having fulfilled an obligation becoming as (or even more) important as actually fulfilling this obligation.

Justice within the organizations	
+	While ambient intelligence may reinforce some human skills and foster people's autonomy, it may also, potentially, increase their power. Ubicomp may lead to a redefinition of power relations and thus result in more egalitarian relationships [16]. As some authors point it out [13, 36], ambient intelligence may be used to integrate people with disabilities in companies. In this perspective, IWEs might be a great tool for positive discrimination.
−	However, the information collected by IWEs may also be used to exclude specific workers (employees that are less productive, people who represent an excessive health risk, etc.) [37, 38].

Justice between organizations	
+	/
−	The costs related to the implementation of ubiquitous systems may create and/or foster inequalities between organizations. Those who cannot afford such systems will not be able to take advantage of the competitiveness they provide.

Performance	
+	As noted above, by reducing the risks associated with accidents and health problems among employees, IWEs should enable companies to increase their performance and help them to be more competitive.
−	In certain circumstances, the costs related to the development of an ubiquitous system may be higher for a company than the potential financial gains due to the improvement of the organizational performance.

Social integration and interactions	
+	Some IWEs promote social relationships within the company, by allowing, for instance, workers who do not work on the same site to be and/or remain in audio–visual contact. In assuming the realization of certain tasks, ubiquitous systems lighten the work of employees. These may take advantage of the gain in time to spend more time to exchange with their colleagues and/or with the client.
−	Although they are supposed to disappear in the background and to strengthen social interaction, intelligent systems also come with some new constraints (alarms, physical distance between people, etc.) that disrupt human relationships.

4 First Normative Considerations

The list of ethical challenges that we have defined is not exhaustive; it deserves to be completed and refined. However, it clearly demonstrates the importance and necessity to carry out reflection on the rationale and relevance of the development of ambient intelligence [1], but also on measures to be taken in order to (ethically and legally) regulate its development in the field of labor.

This aspect has not yet been fully addressed by our research group. However, some proposals have already emerged from our discussions. We present them below in a succinct manner.

4.1 Questioning the Ecosystem Data

Mastery of information is at the heart of the ethical issues raised by the current developments we see in ICTs. As we have seen, IWE is no exception to the rule. Therefore, it seems plausible to begin with a questioning of the normative ecosystem data that is put in place.

In this context, we should particularly ask:

- Who benefits from the data collection?
- Who decides which data are to be collected?
- Who designs the IWE?
- Who drives the system?
- To which purpose or to serve whose interests are the data collected?
- Which data are collected?
- What happens to them?
- Where are they stored?
- How long are they kept?
- How are they destroyed?
- How are changes regarding the reasons (security, control, comfort, etc.) justifying the use of IWE taken into account?
- How are changes or technological developments, which modify the system, taken into account?
- Who is accountable in case of errors occurring during data collection or of their misuse?

These questions hide several ethical issues.

4.2 Respecting Individual Autonomy...

One of the most salient issues is regarding the respect of autonomy and of the balance of power. The decision to implement an IWE is not only the concern of managers:

The workers are also directly involved. Therefore, they should be given the opportunity to participate in the decisions related to the development of such environments. In other words, it is not enough to inform employees about the wish of their employers to implement an IWE. Given the impact ambient intelligence may have on working conditions, it is essential that workers may give (free and informed) consent to the use of such a system and may participate in the reflections on its design (even if no personal data are collected).

This issue does not only arise through the development and implementation of a new IWE. The question of consent must also be raised when changes in the system's functions or in its uses are considered.

4.3 ... and Ensuring the Balance of Power

Moreover, it is essential that the development of an IWE does not result in an imbalanced consideration of the rights and interests of the different stakeholders. In our view, a system that would benefit the employer without providing an at least equal benefit to the employee would be ethically indefensible (in addition to being far more likely to encountering resistance in introduction).

The only "injustice" that we may accept here would imply a positive discrimination, which specifically seeks to redress existing inequalities (disability, illness, etc.). However, even in such cases, the consent and participation of the concerned persons are essential. Confronted with this increasing empowerment of the machine, we have to ensure that users (employees, employers, clients, etc.) are put in a position that allows them to retain control over the system.

4.4 Respecting the Proportionality Principle

The principle of proportionality seems to be another important rule to be respected when collecting and processing data in the context of an IWE. This principle can be understood in several ways.

First of all, *only the information strictly needed to achieve our purpose should be collected.* It is a priori useless to get data related to someone's heartbeat when we wish to ensure that this person does not inhale too many toxic fumes.

In addition, *personal data should be disclosed to third parties only in case of absolute necessity,* and in this case, *the number of persons informed should be reduced to a minimum.* Indeed, to be effective information gathered by an IWE does not need to be made accessible to all (colleagues, employers, etc.). It might be sufficient that the directly concerned person has access to it. And if for high safety reasons it becomes essential to communicate these data to a third party, this communication must be strictly targeted.

Finally, *the length of data retention should be defined in relation with the initially specified needs and goals*, and not in relation to hypothetical future use.

5 Conclusion

Today, the research in Ubicomp is already well advanced and the applications of ambient intelligence in the workplace are arriving on the market. Despite this, reflection on the ethical issues raised by these developments and on ways to address them remains minimal. Through our research, we hope to contribute to this reflection and to stimulate a large open debate with all stakeholders.

References

1. Ehrwein Nihan, C., Kinder-Kurlanda, K.: Ethically intelligent? a framework for exploring human resource management challenges of intelligent working environments. In: van Berlo, A., Hallenborg, K., Corchado, J.M., Tapia, D.I., Novais, P. (eds.) Ambient Intelligence—Software and Applications—4th International Symposium on Ambient Intelligence (ISAmI 2013), pp. 213–219. Springer, Cham/Heidelberg/New York/Dordrecht/London (2013)
2. Dourish, P., Bell, G.: Divining a Digital Future: Mess and Mythology in Ubiquitous Computing. MIT Press, Cambridge MA/London (2011)
3. Ehrwein Nihan, C., Firoben, L., Gonin, F., Hitz, M., Weidmann, J.: Les Technologies Intelligentes un Risque ou une Opportunité pour la GRH?, HR-Vaud, Lausanne (2012)
4. Tavani, H.T.: Ethical aspects of emerging and converging technologies. In: Tavani, T., Ethics and Technology. Controversies, Questions, and Strategies for Ethical Computing, pp. 261–392. Wiley, Hoboken NJ (2011)
5. Wright, D., Gutwirth, S., Friedewald, M., De Hert, P., Langheinrich, M., Moscibroda, A.: Privacy, trust and policy-making: challenges and responses. Comput. Law. Secur. Rev. **25**, 69–83 (2009)
6. Spiekermann, S., Langheinrich, M.: An update on privacy in ubiquitous computing. Pers. Ubiquit. Comput. **13**, 389–390 (2009)
7. Hodel-Widmer, T.B.: Designing databases that enhance people's privacy without hindering organizations. Ethics Inf. Technol. **8**, 3–15 (2006)
8. Duquenoy, P., Burmeister, O.K.: Ethical issues and pervasive computing. In: Godara, V. (ed.) Risk Assessment and Management in Pervasive Computing: Operational, Legal, Ethical, and Financial Perspectives, pp. 263–284. Information Science Reference, Hershey PA (2009)
9. Floridi, L.: The ontological interpretation of informational privacy. Ethics Inf. Technol. **7**, 185–200 (2005)
10. Heesen, J., Siemoneit, O.: Opportunities for privacy and trust in the development of ubiquitous computing. In: Phillips, D., Wiegerling, K. (eds.) Ethical Challenges of Ubiquitous Computing. Int. Rev. Inform. Ethics **8**(12), 47–52 (2007)
11. Bohn, J., Coroamă, V., Langheinrich, M., Mattern, F., Rohs, M.: Allgegenwart und Verschwinden des Computers – Leben in einer Welt smarter Alltagsdinge. In: Grötker, R. (ed.) Privat! Kontrollierte Freiheit in einer Vernetzten Welt, pp. 195–245. Heise-Verlag, Hannover (2003)
12. Hildebrandt, M.: Ambient Intelligence, Criminal Liability and Democracy. Crim. Law Philos. **2**, 163–180 (2008)

13. Bühler, C.: Ambient Intelligence in Working Environments. In: Stephanidis, C. (ed.) Universal Access in HCI, Part II, pp. 143–149. Springer, Berlin (2009)

14. Verbeek, P.-P.: Ambient intelligence and persuasive technology: the blurring boundaries between human and technology. Nanoethics **3**, 231–242 (2009)

15. Kortuem, G., Alford, D., Ball, L., Busby, J., Davies, N., Efstratiou, C., Finney, J., Iszatt White, M., Kinder, K.: Sensor networks or smart artifacts? an exploration of organizational issues of an industrial health and safety monitoring system. In: Krumm, J., Abowd, G.D., Seneviratne, A., Strang, T. (eds.) Ubicomp 2007: Ubiquitous Computing. LNCS 4717, pp. 465–482. Springer, Berlin (2007)

16. Kinder, K.E., Ball, L.J., Busby, J.S.: Ubiquitous technologies, cultural logics and paternalism in industrial workplaces. Poiesis. Praxis: Int. J. Technol. Assessment. Ethics. Sci. **5**(3–4), 265–290 (2008)

17. Langheinrich, M.: Privacy by design—principles of privacy-aware ubiquitous systems. In: Abowd,G.D, Brumitt, B., Shafer, S., (eds.) Ubicomp 2001: Ubiquitous Computing. LNCS 2201, pp 273–291. Springer, Berlin (2001)

18. Corchado, J.M., Bajo, J., de Paz, Y., Tapia, D.I.: Intelligent environment for monitoring Alzheimer patients, agent technology for health care. Decis. Support Syst. **44**, 382–396 (2008)

19. Bellotti, V., Edwards, K.: Intelligibility and accountability: human considerations in context-aware systems. Hum. Comp. Interact **16**, 193–212 (2001)

20. Duan, Y., Canny, J.: Protecting user data in ubiquitous computing: towards trustworthy environments. In: Martin, D., Serjantov, A. (eds.) Privacy Enhancing Technologies. LNCS 3424, pp. 167–185 (2005)

21. Boos, D., Guenter, H., Grote, G., Kinder, K.: Controllable accountabilities: the internet of things and its challenges for organisations. Behav. Inf. Technol. **32**(5), 449–467 (2013)

22. Spiekermann, S., Pallas, F.: Technology paternalism—wider implications of ubiquitous computing. Poiesis, Praxis: Int. J. Technol. Assessment. Ethics. Sci. **4**(1), 6–18 (2006)

23. Wieland, M., Längerer, C., Leymann, F., Siemoneit, O., Hubig, C.: Methods for conserving privacy in workflow controlled smart environments. a technical and philosophical enquiry into human-oriented system design of ubiquitous work environments. In: Third International Conference on Mobile Ubiquitous Computing, Systems, Services and Technologies, pp. 16–21. IEEE Computer Society, Washington DC (2009)

24. Ratto, M.: Ethics of seamless infrastructures: resources and future directions. In: Phillips, D., Wiegerling, K. (eds.) Ethical Challenges of Ubiquitous Computing. Int. Rev. Inform. Ethics **8** (12), 20–27 (2007)

25. Ehrwein Nihan, C.: Penser la crise ou plaidoyer pour une réflexion critique sur la crise de la crise à partir de l'œuvre d'Hannah Arendt. In: Müller, D., Waterlot, G. (dir.), Revue d'Ethique et de Théologie Morale **276**, 43–60 (2013)

26. Santos, M.S., Pitt, J.: Ubiquitous computing and pervasive adaptation of social norms in workplace design. In: Proceedings of the Symposium on Mental States, Emotions and their Embodiment, AISB 2009 Convention, pp. 32–35. Society for the Study of Artificial Intelligence and the Simulation of Behaviour (2009)

27. Weiser, M., Gold, R., Brown, J.S.: The origins of ubiquitous computing research at PARC in the late 1980s. IBM Syst. J. **38**(4), 693–696 (1999)

28. Morris, M., Lundell, J., Dishman, E.: Catalyzing social interaction with ubiquitous computing: a needs assessment of elders coping with cognitive decline. In: Proceeding CHI EA '04 CHI '04 Extended Abstracts on Human Factors in Computing Systems, pp. 1151–1154. Association for Computing Machinery (ACM) (2004)

29. Hilty, L.M., Som, C., Köhler, A.: Assessing the human, social, and environmental risks of pervasive computing. Hum. Ecol. Risk Assess **10**(5), 853–874 (2004)

30. Tang, J.C.: Ubiquitous Computing: Individual Productivity at the Expense of Social Good. IBM Research, San Jose CA (2005). Paper available online: http://www.vs.inf.ethz.ch/events/ubisoc2005/UbiSoc%202005%20submissions/04-Tang-John-NEW.pdf Accessed 21 April 2014

31. Curry, M.R.: Being there then: ubiquitous computing and the anxiety of reference. In: Phillips, D., Wiegerling, K. (eds.) Ethical Challenges of Ubiquitous Computing, Int. Rev. Inform. Ethics **8**(12), 13–19 (2007)
32. Foucault, M.: Surveiller et Punir. Naissance de la Prison. Gallimard, Paris (1975)
33. Floridi, L. (ed): The Cambridge Handbook of Information and Computer Ethics. Cambridge University Press, Cambridge/New York/Melbourne/Madrid7Cape Town/Singapore (2010)
34. Hilty, L.M., Behrendt, S., Binswanger, M., Bruinink, A., Erdmann, L., Fröhlich, J., Köhler, A., Kuster, N., Som, C., Würtenberger, F.: Das Vorsorgeprinzip in der Informationsgesellschaft. Auswirkungen des Pervasive Computing auf Gesundheit und Umwelt. TA-Swiss 46, Berne (2003)
35. Boos, D., Grote, G.: Designing controllable accountabilities of future internet of things applications. Scand. J. Inf. Syst. **24**(1), 3–28 (2012)
36. Villarubia, G., Sánchez, A., Barri; I., Rubión, E., Fernández, A., Rebate, C. Cabo, J.A., Álamos, T., Sanz, J., Seco, J., Zato, C., Bajo, J., Rodríguez, S., Corchado, J.M.: Proximity Detection Prototype Adapted to a Work Environment. In: Novais, P., Hallenborg, K., Tapia, D.I., Rodríguez, J.M.C. (eds.) Ambient Intelligence–Software and Applications. 3rd International Symposium on Ambient Intelligence (ISAmI 2012), Advances in Intelligent and Soft Computing Vol. 153, pp. 51–58. Springer, Berlin/Heidelberg (2012)
37. Gandy, O.H.: Engaging rational discrimination: exploring reasons for placing regulatory constraints on decision support systems. Ethics Inf. Technol. **12**, 29–42 (2010)
38. Ehrwein Nihan, C.: Intelligent Working Environments, Handling of Medical Data and the Ethics of Human Resources. In: Omatu, S., De Paz Santana, J.F., Rodríguez González, S., Molina, J.M., Bernardos, A.M., Corchado Rodríguez, J.M. (eds) Distributed Computing and Artificial Intelligence. 9th International Conference, Advances in Intelligent and Soft Computing Vol. 151, pp. 429–436. Springer, Berlin (2012)

Stakes of Ubicomp-ICT in the Light of an Ethico-political Standpoint

Hugues Poltier

Abstract This paper discusses the issues involved in ICT from an ethico-political perspective, which states that an innovation is fully justifiable when it contributes to the empowerment, equality, and autonomy of the agent targeted. With this criterion in mind, discussing the foreseeable impacts of ubicomp-ICT, it focuses on the dangers linked to the enhanced asymmetry in the distribution of power within the organization which it is likely to bring about. If left to "laissez-faire", it is finally suggested, it might even be seen as a threat to democracy.

Keywords Ethico-political standpoint · Power asymmetry · Worse-off vulnerability

1 Introduction

The papers collected in this dossier indicate some of the ongoing transformations of life-world through the design and development of new applications based on ICT and more widely on ubicomp-ICT. In particular, they show how a well-designed complex system of collecting and treating data relevant to a given practical field may enhance the monitoring capacities of the persons responsible for its good and efficient functioning—and, as a result, the good of the final receivers who have been, since the beginning, the end goal of the whole process. Their benefits are relatively easy to pinpoint, provided they are designed with the concern to take into account all the relevant aspects involved in the practice, i.e., not only the concern for a greater efficiency from the manager's point of view, but also the concerns of the patient or of the agent (the worker or the disabled) involved in the process (De Paz, Rodriguez, Zato and Corchado, Voirin). These benefits range from greater cost-efficiency (allowing a monitoring far more precise and more responsive to immediate problems or troubles arising with less staff), to greater accountability of a company toward its clients or insurances (Kinder-Kurlanda and Boos), via the

H. Poltier (✉)
University of Lausanne, Lausanne, Switzerland
e-mail: hugues.poltier@unil.ch

© Springer International Publishing Switzerland 2015
K. Kinder-Kurlanda and C. Ehrwein Nihan (eds.), *Ubiquitous Computing in the Workplace*, Advances in Intelligent Systems and Computing 333,
DOI 10.1007/978-3-319-13452-9_7

integration of disabled people in the production process. These studies show that ubicomp technologies are part of the ongoing process of developing technology-based devices designed to improve and facilitate the realization of tasks while increasing the safety of the people involved.

1.1 Preliminary Steps

A conclusion toward which these contributions as well as others converge is the almost infinite versatility of ICT. This means that there is no foreseeable limit to what can be digitized. Virtually any practical field is de jure captured in a process of digitization so that it can be decomposed into a set of manipulable items and data which can be stored, reorganized for a process, monitored through an adequate interface, automatized, reconfigured, subject to an a posteriori control, etc.

This dossier involves contributions that address some of the ethical issues at stake. Ehrwein Nihan and Hilty offer a large survey of some of the more pressing ethical questions raised by these new technologies while Wiegerling discusses their impacts on our life-world ("Wirklichkeit"). Assuming these discussions are known to the reader, my paper will focus on what I will call questions from an ethico-political standpoint, less represented in this dossier.

2 An Ethico-political Perspective

Political ethics has to do basically with distribution of power and goods within society. The normative basic principle stems from the democratic idea, namely that the form and content of the social life and activities are ultimately to be decided and governed by the people in the interest of the people—each having an equal share in the decision making and having an equal right in the good produced. From this standpoint, a novelty ought to be assessed according to its contribution to the self-empowerment of the people—assuming that the demos acts according to its very principle as long as it ensures and promotes equal rights and power of all its members, any deviation from this principle suggesting a possible corruption and deviation toward a form of tyranny, however soft it may be.

We can spell out the triangle constitutive of this political ethics: to be fully in accordance with its principles an innovation has to satisfy the requirements of equality, liberty, and autonomy. The agent whose benefit is the end goal of the device has to benefit in empowerment: he/she must be able to perform acts that were not available to him/her without the device, sure. But not only. He/she has to experience also a sense of control over his/her life, of having a say on issues pertaining, not only to his/her personal life, but also to collective stakes which impact his/her life. Shortly, the beneficiaries must, in some way, be involved in the designing process. Or else: they must be treated as "subjects", not only as objects of the process. It is not only about enjoying, but also about participating. Furthermore, this participation should not limit itself to assessing the immediate personal benefit, but also the long-term effects of the proposed innovation: there ought not to be a full focus solely on the immediate personal benefit. Equality is satisfied if and only if one has a sense that one's voice in the decision making is recognized as having a weight and one is not treated solely as being defined by one's personal and short-term interests. Otherwise, one is just being treated as an object, a living object whose well-being is paternalistically taken care of.

So, our question, in what follows, circle around an effort to appreciate ubicomp-ICT in terms of its likely impacts on us as political stakeholders in the building of a collective of citizens where everyone takes part, as a full political subject, in the collective decision making. More precisely: does the ICT satisfy or even promote these ethico-political requirements? Does it threaten them in some way or another? Are the possible threats dealt with? In the way ubicomp is actually developing, are these stakes clearly identified and taken care of?, etc.

3 A Process in Limbo

Before addressing this issue, let me observe that ubicomp-ICT is still in limbo. It is hardly disputable that it will penetrate more and more, if not all, at the very least all the social and professional environments in which an ongoing and pervasive monitoring is likely to bring greater cost-efficiency and safety (from health hazards, bad conduct, terrorist threat, etc.). The reason for this highly likely trend is relatively simple to formulate: for a person in charge, it is unlikely not to embrace ways that will improve his/her monitoring power and efficiency. Being accountable, they are supposed to take all measures, once available (or possibly so with a given investment), which one can reasonably foresee will foster better monitoring and cost-effectiveness of a given process. Being accountable to the board of directors, and beyond, the shareholders, they are under an obligation of "results". This involves the obligation to develop all possible new devices based on the most recent and tested new technologies so as to ensure the competitiveness of the firm or the avoidance of resource-wasting in the public sector (given the well-known pervasive pressure on the public budget in contemporary politics). Not complying with this requirement puts the manager at risk, the risk to be fired and replaced by somebody more prone to satisfy the board's or shareholders expectations.

4 Ubicomp: Toward a Reshaping of the Balance of Power?

Assuming this assessment, the question to be asked, from an ethico-political standpoint, is as follows: how is ubicomp-ICT to be expected to reshape relationships throughout society? And this concerns all levels: at the level of society to its members as well as of the individual relating to other individuals, associations, firms, insurances, intermediate collectivities, state authorities, or society as such. The question may also be asked at a societal level (close to Wiegerling's approach) of how it will affect our sense of togetherness and life-world. My perspective, nevertheless, is more focused on the dimension of power and its distribution in society: how is it to be expected that ubicomp-ICT will affect the distribution of control—and hence power—in society?

With this issue in mind, I deem it important to stress a dimension insufficiently discussed in this dossier. One largely shared view in all the discussions on ICT has been its disseminating and therefore individual-empowering potential. Through his/her mastering of some ICT techniques, the individual gains power to reshape his/her life and course of action. He/she can create new possibilities and powers in interacting with the world. This is not disputable. But this constitutes only one side of the coin, as suggests the recent discussion on the Big Data issue which has gained momentum through the whole NSA affair. This recent scandal has raised concerns about the threat against democracy posed by these formidable concentrations of data. It was as if, suddenly, we realized the power represented by very large amounts of data concentrated under the control of one entity.

This last notation can be related to some points mentioned in the dossier, in particular, in Kinder-Kurlanda and Boos's paper. In the cases they report and comment, it can be clearly noticed that the requirements for better transparency and tracing of the liability go downwards: the top of the organization wants to have a better monitoring of all the course of action so as to check and, when appropriate, improve cost-effectiveness and security. This means that the lower levels of the organization tend to be transparent toward the upper ones, but not the reverse. Furthermore, it can be noted, in the liability cases, that the request comes from outside the company, namely the insurances who, before footing the bill, want a complete information on the circumstances having caused a given damage so as to determine whether a faulty behavior cannot be imputed to the company or the agent involved, justifying a penalty to the one or the other. As the final payer, the insurance has been recognized as enjoying a legal right to check the conformity of behavior to regulations at all the levels involved in the realization of the action. And now, with the generalization of ubicomp, this is a request that is more and more possible to honor—and therefore tends to get compulsory. In effect, it is more and more accepted that the counterpart of their obligations should be a right to a thorough monitoring of the action of the whole chain of subordinates or insured with the help of the most recent and powerful technologies. The monitored parties, in their eyes, are under the obligation to endure this examination of their practice for the sake of cost-efficiency and safety. Any reluctance from the monitored ones to accept this procedure would be seen as suspicious and devastating for future collaboration.

As a consequence of the power to demand the fullest possible account of the actions on the ground, the lower staff are the object of a control stronger than ever. In effect, it is arguable that they largely lose in autonomy and participation in decision making since (1) the decision regarding the norms to comply with are a privilege of the top of the collective structure contributing to the realization of the good and (2) all their movements, activities, time-responses, etc., are objects of the monitoring through GPS and other sensors transmitting information to a base apt to follow in real time their location, gestures, procedure following, etc. In a way, one can deem this positive: the management can ensure by these devices that no time (i.e., money) is wasted by the employees, no unsafe procedure engaged in, no rule breaking behavior committed, etc. Undoubtedly, (provided the work plan and the rules have been well thought over by the management) it results in a more cost-effective way of dealing with the tasks as well as in a more rule-compliant behavior of the staff (causing, in principle, less possible damages, and therefore less costs). This can be seen as the positive side of the coin.

Considered from the lower end, this whole process may well be experienced in another far less sunny light. Since there is the presumption that these devices allow an ongoing and strictly individual follow-up, it means a control leaving less and less—and once fully operational, none at all—room to the individual worker in the execution of his/her tasks. Or, to put it in other words: the more individualized the control, the more direct the power of the hierarchy on each individual, the more impossible for the employee to evade the demands stemming from the hierarchy; the more asymmetrical the relationships in the organization. This trend can only be increased by the demands of a greater efficiency stemming from the shareholders', deputies in the name of their duty to protect and promote the owners' interests.

This can be described in another way: by virtue of the organization of the information flow in the corporation, a given individual is entitled to gather information from subordinates only and is supposed to transmit all relevant information to his/her line manager while the latter transmits downwards orders and prescriptions. As a subordinate, I have no right to be informed on what is going on in the firm, whereas, as a trend, the top management has a right to know everything, at least all that is relevant to the management of the company. The right of control of the latter on the former (and now, not only on the workplace: think of Facebook and all the social networks as tools allowing the company a control largely exceeding the sole behavior on the workplace. With the latest developments of ubicomp, it is less and less clear whether the employee's claim to privacy is still a reasonable one) is potentially total, whereas the reverse is close to nil.

Briefly then, the situation can be summarized through the following features: asymmetry, collection, and concentration of individualized data at the top of the organization (asymmetrical transparency), control following a top-down path, increased control of each individual in the hierarchy by the top management; the top is not accountable to the lower levels, only to the ones above them, i.e., ultimately (in stock companies) the board of directors and the shareholders assembly.

So as to prevent any misunderstanding, let me be clear about what has been said. The problem is not control as such. Since we share a world, we have a right to expect that our fellow citizens do not commit given actions; and to ensure that, it is arguable that certain controls be enforced—included on us.

The problem lies in the asymmetry of controls, especially in the workplace. Some, by virtue of their property, are in a position to require and obtain a right to an extensive control on all the ones who contribute to the generation of value of the firm they own. Seen in this light, ubicomp is, in and of itself, not a new ethical issue. Rather, it constitutes a resource in the hands of management allowing them to refine and make more pervasive the tracking of the collaborators of a firm. And given the balance of power, the outcome resulting from the existence of these new devices is the aggravation of the former asymmetry, diminishing the "invisibility zone" of the lower ones, hence their ability to resist the increasing demands of their employer.

This asymmetry is still sharpened by two reinforcing factors which, in themselves, are foreign to the examined field: the increasing power of multinational companies that are less and less controlled by the political; the high rate of unemployment. The combination of these two factors aggravates the vulnerability of the employed in that it diminishes their collective bargaining power, making them more dependent upon their employer's pleasure.

I once again want to make sure not to be misunderstood. The point I am making is very simple indeed: a technical device is a power, in the broad sense of the term. As such, its control is always a stake: the problem is not the power as such but the question of knowing who have control over it, whether control is evenly distributed among the stakeholders or concentrated in the hands of the shareholders. In the former case, one can argue that no power gained through technological innovation poses as such an ethico-political threat; no doubt, the question of its opportunity, beneficiary character related to its costs and environmental impacts remains open, but this is another issue. In the latter case, one has to say that the more control possibility a technique brings, the more dangerous it is from an ethico-political standpoint, and the greater is the threat to the democratic ideal of a community of free and equal citizens. The more possibility of control those at the top have over those at the bottom, the more extensive is their power to enforce compliance with their expectations and norms. Being absolutely transparent to the line management in terms of performance, norms compliance, commitment to the achievement of the firm's objectives, etc., the employees are exposed to a very large restriction of their autonomy in their mission fulfillment. What adds up to this threat is the increasing ability multinational companies have to evade serious controls of the political. In case of a breach of the law from such a firm, the likelihood to get exposed to a sanction involving a threat to the company's existence gets vaguer than ever. In effect, these societies tend to treat the possible sanctions as a mere financial risk which they take into account in their planning: there is no such thing as a penal condemnation of a corporation, however severe may be the damage inflicted on persons, on their liberty and autonomy, etc.

5 To Conclude

In and of themselves, from an ethico-political standpoint, the new possibilities offered by ubicomp-ICT pose no problem. They can be appreciated as belonging to the long history of techniques seen as the many innovative ways we, as a collective, have designed in order to improve the fulfillment of needs, and therefore of the common good. The ethico-political standpoint adopted in this paper leaves open the full appreciation of a given innovation, regarding the adequacy of the designed device, for instance regarding its efficacy, efficiency, and environmental impact. The central issue, from our standpoint, is the effect in terms of the distribution of power within the society. With this issue in mind, our conclusion is that, if concentrated in the shareholders' hands, these techniques, offering a controlling power greater than ever in the past, we, as citizens collective, should be extremely careful not to leave at their discretion the power to use these new controlling tools' powers. We should reclaim our right to have control over our lives.

Retraction Note to: An Integrated System for Helping Disabled and Dependent People: AGALZ, AZTECA, and MOVI-MAS Projects

Juan F. De Paz, Sara Rodríguez, Carolina Zato,
and Juan M. Corchado

Retraction Note to:
Chapter 1 in: K. Kinder-Kurlanda and C. Ehrwein Nihan (eds.), *Ubiquitous Computing in the Workplace*, **Advances in Intelligent Systems and Computing 333, https://doi.org/10.1007/978-3-319-13452-9_1**

The Series Editor and the publisher have retracted this chapter. An investigation by the publisher found that a number of chapters, including this one, from multiple conference proceedings raise various concerns, including but not limited to inappropriate or unusual citation behavior and undisclosed competing interests. Based on the findings of the investigation, the Series Editor and the publisher no longer have confidence in the results and conclusions of this chapter.

Authors Sara Rodríguez, Juan M. Corchado and Carolina Zato disagree with this retraction. Author Juan F. De Paz have not responded to correspondence regarding this retraction.

The retracted version of this chapter can be found at
https://doi.org/10.1007/978-3-319-13452-9_1

© Springer International Publishing Switzerland 2024
K. Kinder-Kurlanda and C. Ehrwein Nihan (eds.), *Ubiquitous Computing in the Workplace*, Advances in Intelligent Systems and Computing 333,
https://doi.org/10.1007/978-3-319-13452-9_8

Author Index

© Springer International Publishing Switzerland 2015
K. Kinder-Kurlanda and C. Ehrwein Nihan (eds.), *Ubiquitous Computing in the Workplace*, Advances in Intelligent Systems and Computing 333,
DOI 10.1007/978-3-319-13452-9

Printed in the United States
by Baker & Taylor Publisher Services